EPA 820-R-13-001

Biological Assessment Program Review:

Assessing Level of Technical Rigor
to Support Water Quality Management

February 2013

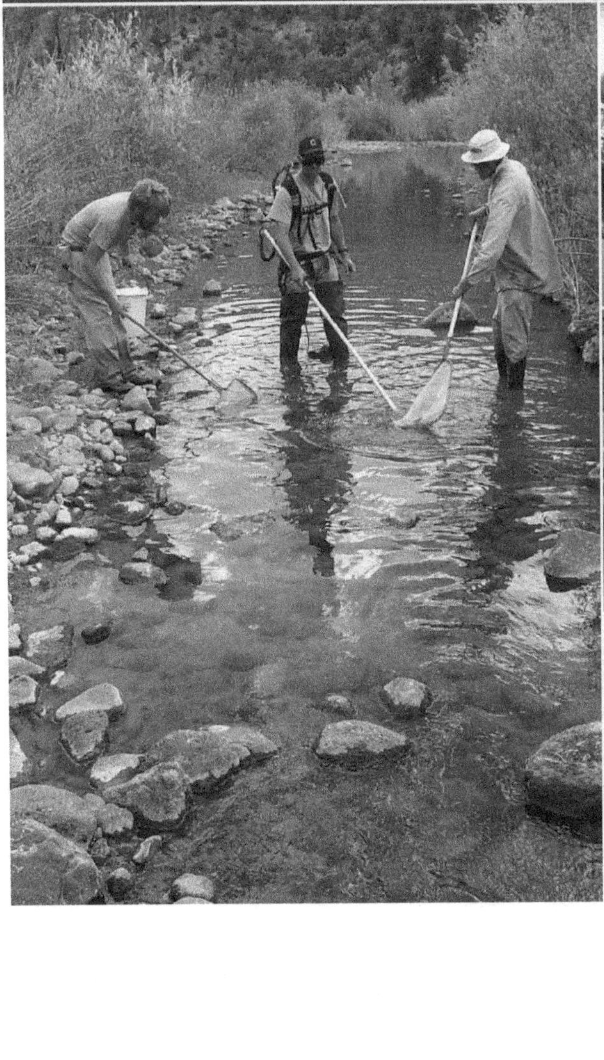

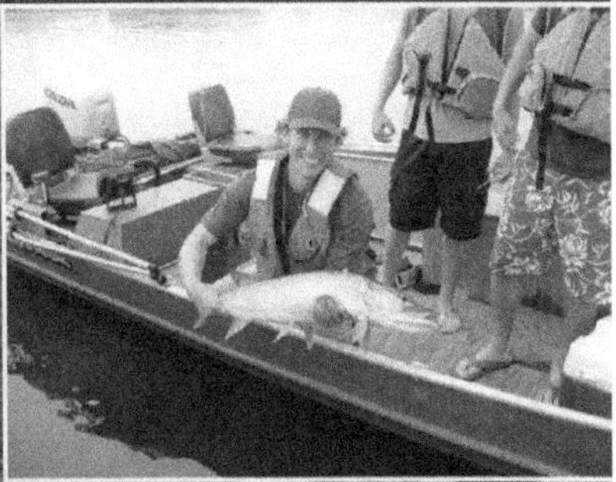

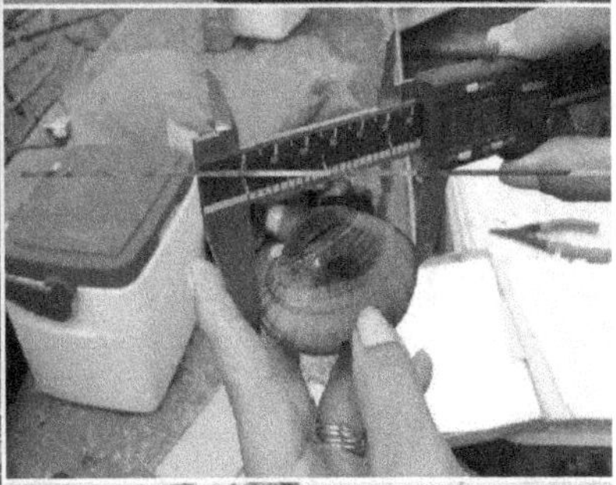

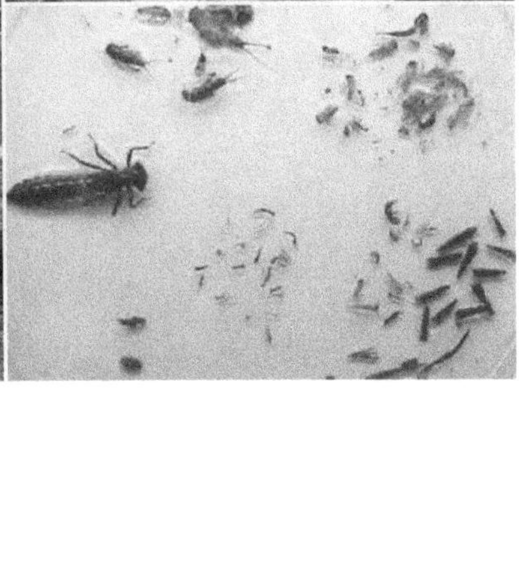

Front cover sources:
1. EPA; 2. USGS; 3. USGS; 4. EPA

United States	Office of Science and Technology	February 2013
Environmental Protection	Washington, DC 20460	EPA 820-R-13-001
Agency		

Biological Assessment Program Review:

Assessing Level of Technical Rigor to Support

Water Quality Management

Disclaimer

The discussion in this document is intended to provide information on advancements in the field of biological assessments and on use of biological assessments to support state water quality management programs. The statutory provisions and the U.S. Environmental Protection Agency (EPA) regulations described in this document contain legally binding requirements. This document is not a regulation itself, nor does it change or substitute for those provisions or regulations. The document does not substitute for the Clean Water Act (CWA) or EPA or state regulations. Thus, it does not impose legally binding requirements on EPA, states, tribes, or the regulatory community. This document does not confer legal rights or impose legal obligations on any member of the public.

While EPA has made every effort to ensure the accuracy of the discussion in this document, the obligations of the regulated community are determined by statutes, regulations, and other legally binding requirements. In the event of a conflict between the discussion in this document and any statute or regulation, this document will not be controlling.

The general descriptions provided here might not apply to a situation depending on the circumstances. Interested parties are free to raise questions and objections about the substance of this document and the appropriateness of the application of the information presented to a situation. This document does not make any judgment regarding any specific data gathered or determinations made by a state or tribal biological assessment program or the use of such data in the context of implementing CWA programs. Mention of any trade names, products, or services is not and should not be interpreted as conveying official EPA approval, endorsement, or recommendation.

This is a living document and might be revised periodically. EPA could revise this document without public notice to reflect changes in EPA policy, guidance, and advancements in field of biological assessments. EPA welcomes public input on this document at any time. Send comments to Susan Jackson, Office of Science and Technology, Office of Water, U.S. Environmental Protection Agency, 1200 Pennsylvania Avenue, Mail Code 4304T, Washington, DC 20460.

BIOLOGICAL ASSESSMENT PROGRAM REVIEW:

ASSESSING LEVEL OF TECHNICAL RIGOR TO SUPPORT

WATER QUALITY MANAGEMENT

Contact Information

For more information, questions, or comments about this document, please contact Susan Jackson, U.S. Environmental Protection Agency, at Office of Science and Technology, Office of Water, U.S. Environmental Protection Agency, 1200 Pennsylvania Avenue, Mail Code 4304T, Washington, DC 20460, or by email at jackson.susank@epa.gov.

Acknowledgments

Thank you to the following state and tribal agencies for their support with developing and piloting the biological assessment program review process:

Alabama Department of Environmental Management
Arizona Department of Environmental Quality
California State Water Resources Control Board
Connecticut Department of Environmental Protection
Colorado Department of Public Health and Environmental Management
Florida Department of Environmental Protection
Fort Peck Tribes (Assiniboine and Sioux)
Illinois Environmental Protection Agency
Indiana Department of Environmental Management
Iowa Department of Natural Resources
Maine Department of Environmental Protection
Massachusetts Department of Environmental Protection
Michigan Department of Environmental Quality
Minnesota Pollution Control Agency
Missouri Department of Natural Resources
Montana Department of Environmental Quality
New Hampshire Department of Environmental Services
New Mexico Environment Department
Ohio Environmental Protection Agency
Rhode Island Department of Environmental Management
Texas Commission on Environmental Quality
Vermont Department of Environmental Conservation
Wisconsin Department of Natural Resources

Acknowledgments (cont.)

Thank you to the following scientists for their support and, for several, their leadership in developing the biological assessment program review process:

Chris Yoder, Midwest Biodiversity Institute
Susan Davies, Maine Department of Environmental Protection

Lester Yuan, Janice Alers-Garcia, Thomas Gardner, EPA Office of Science and Technology
Susan Holdsworth, Sarah Lehmann, Ellen Tarquinio, Treda Grayson,
EPA Office of Wetlands, Oceans, and Watersheds
Wayne Davis, Office of Environmental Information

Stephen Silva, Katrina Kipp, Jennie Bridge,
Hilary Snook, Diane Switzer, EPA Region 1
Linda Holst, Edward Hammer, EPA Region 5
Charlie Howell, Michael Schaub, EPA Region 6
Gary Welker, Catherine Wooster-Brown, EPA Region 7
Tina Laidlaw, EPA Region 8
Terrance Fleming, EPA Region 9

David Peck, Steve Paulsen, John Stoddard, James Lazorchak, Scot Hagerthey,
Carolina Penalva-Arana, Giancarlo Cicchetti,
EPA Office of Research and Development

Charles Hawkins, Bernard Sweeney, Society of
Freshwater Science Taxonomic Certification Committee

Michael Barbour, Jeroen Gerritsen, Clair Meehan, Christoph Quasney,
Jennifer Stamp, James Stribling, Tetra Tech, Inc.

Contents

Figures

Tables

Foreword

State and tribal water quality agencies face challenges to ensure that the best available science serves as the backbone of their monitoring and assessment programs. The degree of confidence with which biological assessment information can be used to answer water quality management questions relies to a considerable degree on a program's level of technical rigor.

This document provides a process, including materials, for states and tribes to evaluate the technical rigor and breadth of capabilities of a biological assessment program. The review is intended to help states and tribes answer the following questions:

- What are the strengths of my technical program?
- What are the limitations of my technical program?
- How do I determine priorities and allocate resources to further develop the technical capabilities of my existing program?
- If I want to use biological assessments to more precisely define my designated aquatic life uses and develop numeric biological criteria, how do I begin technical development?

Using the program review process described in this document, states and tribes can identify the technical capabilities and the limitations of their biological assessment programs and develop a plan to build on the program strengths and address the limitations. The U.S. Environmental Protection Agency (EPA) recommends that the review include both EPA regional participants and agency program managers and staff, and that it be facilitated by a technical expert with expertise in biological assessments and biological criteria derivation. As part of the review process, a state or tribe evaluates how it currently uses biological assessment information to support its overall water quality management program and considers potential future applications using information gained by a strengthened technical program.

The document includes a description of 13 technical elements of a biological assessment program, provides a checklist for evaluating the level of technical development for each element, and includes a method for characterizing the overall level of program rigor. As a technical program is improved, biological assessment information can be used with increasing confidence to support multiple water quality program needs for information. Such needs include more precisely defined aquatic life uses and approaches for deriving biological criteria, monitoring biological condition, supporting causal analysis, and developing stressor-response relationships.

This document is intended to be used as a "how to" manual to guide technical development of a biological assessment program for providing information to meet multiple water quality information needs. Water quality agencies can use the outcomes of the programmatic review to develop the technical strengths of their biological assessment programs and allocate resources to build as robust programs as their resources will allow. The highest level of technical development as described in this document can be thought of as a well-equipped toolbox. Not all tools need to be applied all the time and in all situations. For a water quality program, the type and level of quality of a biological assessment tool (e.g., a collection method, monitoring design, or analytical approach) will depend on the question being asked and the specific environmental circumstances. For this reason this document does not, and is not intended to, establish minimum expectations regarding the amounts or types of biological data that might be considered necessary in the context of decision making in Clean Water Act regulatory programs. However, understanding the different programmatic expectations for the biological assessment data guides the technical review and recommendations for technical development.

CHAPTER 1: BIOLOGICAL ASSESSMENT PROGRAM EVALUATION

1.1 Background

A biological assessment is an evaluation of the biological condition of a water body using surveys of the structure and function of resident biota, including migratory biota that reside in the water body for at least one part of their life cycle (USEPA 2011b). Biological assessment information is important to effectively and accurately answer water quality management questions about condition, protection, and restoration. It is a principal monitoring tool for state and tribal water quality agencies

(referred to throughout as water quality agencies) and is used to varying degrees and purposes by all 50 states and increasingly by tribes (USEPA 2002b, 2011c). Over the past 20 years, water quality agencies have developed different abilities to use biological assessment information for water quality management. An agency's ability to use this information at the appropriate level of precision and accuracy to answer a given management question is called its *technical capability*. The technical capability of a program is dependent on its level of *technical rigor*. For the purposes of this document, a technically rigorous biological assessment program:

- Uses scientifically accepted and documented methods.

- Adheres to methods and protocols.

- Documents quality assurance and quality control.

- Provides information to support multiple WQM programs.

1.2 Why Is the Level of Technical Rigor Important?

The technical rigor of a biological assessment program determines the degree of accuracy and precision in assessing biological condition and deriving stressor-response relationships. With increasing technical rigor, a water quality agency gains increased confidence in data analysis and interpretation, as well as more comprehensive support for a variety of water quality management activities, including the following:

- More precisely defining goals for aquatic life use protection.

- Deriving biological criteria.

- Identifying high quality waters and establishing biological condition baselines.
- Identifying waters that fail to support designated aquatic life uses.
- Supporting development of water quality criteria.
- Conducting causal analysis.
- Monitoring biological response to management actions.

 This document is intended to be used as a road map for technical development of a biological assessment program. It provides a step-by-step process for evaluating both the technical rigor of a water quality agency's biological assessment program and the extent to which the water quality agency uses the information to support overall water quality management. The evaluation is based on the degree of technical development of the biological assessment program's survey design, methods, analysis, and interpretation; how biological assessments are integrated into and supported by the monitoring program; how the agency currently uses biological assessments to support its water quality programs; and how it intends to use biological assessments in the future.

The end goal of this evaluation process is an action plan for technical program development and recommendations to enhance the use of biological assessments to support the agency's overall water quality management program (USEPA 2011c). The plan specifies incremental steps for technical and program development based on the strengths and gaps identified in the evaluation.

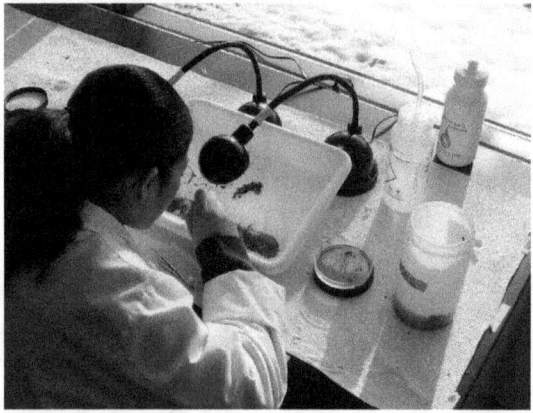

To date, this process has been applied to biological assessment programs for river and streams and reviews conducted with 22 states and 1 tribe (Yoder and Barbour 2009). However, the technical elements and the review process are applicable to other water body types with water body-specific modifications for biological assessment design, methods, and data analysis.

1.3 The Technical Foundation for a Biological Assessment Program

The determination of a biological assessment program's level of technical rigor is on the basis of evaluating 13 technical elements that provide the foundation of its biological assessment design, data collection and compilation, and analysis and interpretation (Figure 1-1). *Biological assessment design* includes temporal and spatial considerations in developing a monitoring program and selection of sampling sites, characterizing and accounting for natural variability, and determining reference condition. *Data collection and compilation* includes field and laboratory protocols and data handling, typically included in agency standard operating procedures (SOPs). *Analysis and interpretation* comprise all of the data analysis, interpretation, and review procedures used after data are obtained. The 13 technical elements are based on U.S. Environmental Protection Agency's (EPA's) Consolidated Assessment and Listing Methodology (CALM) guidance on collection and use of water quality data and information for environmental decision making (USEPA 2002a), and on EPA's *Evaluation Guidelines for Ecological Indicators* (Jackson et al. 2000; Kurtz et al. 2001). The evaluation guidelines described 15 guidelines in 4 areas (termed "phases" in the Guidelines) comprising conceptual relevance of the indicator, feasibility of implementation, response variability, and interpretation and utility. The CALM guidance describes seven critical technical elements of a biological assessment program. In that guidance EPA also describes four levels of technical program rigor, Levels 1 through 4, with Level 4 being the highest level of rigor. As described in chapter 2 of this document, the original 7 critical technical elements have been refined and expanded to 13 elements on the basis of a water quality agency's assessment program reviews conducted beginning in 2004 (Yoder and Barbour 2009; USEPA 2010b).

Technical Elements of a Biological Assessment Program

Biological assessment design
1. Index period
2. Spatial sampling design
3. Natural variability
4. Reference site selection
5. Reference conditions

Data collection and compilation
6. Taxa and taxonomic resolution
7. Sample collection
8. Sample processing
9. Data management

Analysis and interpretation
10. Ecological attributes
11. Discriminatory capacity
12. Stressor association
13. Professional review

Figure 1-1.The critical technical elements.

The technical elements and the level of development for a rigorous biological assessment program are discussed in more detail in chapter 2. Assessment of the technical elements is the technical backbone of the program review process, and it provides the detailed information needed by an agency program to develop its technical program. An estimate of overall level of program rigor is assigned based on the scoring of the technical elements that correspond with a program's increasing ability to detect incremental levels of biological change along a gradient of stress, associate biological response to stressors and their sources, and integrate biological assessments with other environmental data and information.

1.4 The Biological Program Review Process

The biological program review is a systematic *process* to evaluate the technical rigor of a water quality agency's biological assessment program and to identify logical next steps for overall program improvement. The review is typically conducted over two to three days for both a thorough evaluation of the technical elements and for agency cross-program discussions on the use of biological assessment data and information to support the overall water quality management program. The purpose of the cross-program discussions is to provide an opportunity for managers and staff from different water quality programs to identify the type and level of rigor of biological assessment that best addresses their information needs. Additionally, personnel can share their needs and timing for information to optimize collection and delivery of the data. These discussions might reveal areas for program improvement and coordination that will foster more efficient and comprehensive application of biological assessments. An improved understanding of how an agency uses biological assessment information in its water quality programs helps answer the "so what" question for why an agency would allocate staff and resources for technical development.

The review includes both EPA regional participants and agency program managers and staff, and it is typically facilitated by an independent technical expert with expertise in biological assessments and in biological criteria derivation.

The review team first evaluates the 13 technical elements of a biological assessment program. Each technical element receives a score on the basis of its current state of technical development. These scores are then summed for an overall program score—a higher score reflecting a higher level of technical development, corresponding with increased capability and confidence in use of biological assessment data.[1] A Level 4 assignment is the highest ranking, and Level 1 is the lowest ranking. These levels reflect sequential stages in technical development of a biological assessment program and are intended as a guide for assessing progress and targeting resources.

The review process is designed to evaluate the key gaps in a technical program and to identify incremental steps for addressing the gaps. The scoring of the individual elements provides the essential information for identifying these technical gaps. Incremental improvements in the individual technical elements are followed, often in a short time, by corresponding improvements in the technical capability of the overall program (Figure 1-2). At all levels of technical development described in this document, a state or tribal program is able to use biological assessment information to carry out Clean Water Act (CWA) activities. For example, a defensible decision that aquatic life use is impaired can be based on a qualitative visual observation of overwhelming biological evidence such as nearly total dominance of pollutant

---

[1] Because the overall score is the result of the summation of individual scores for the 13 separate elements, the overall score does not establish minimum expectations regarding a state's ability to make decisions in context of different CWA regulatory programs. At all levels of technical development, biological assessment information can be used to support water quality decisions.

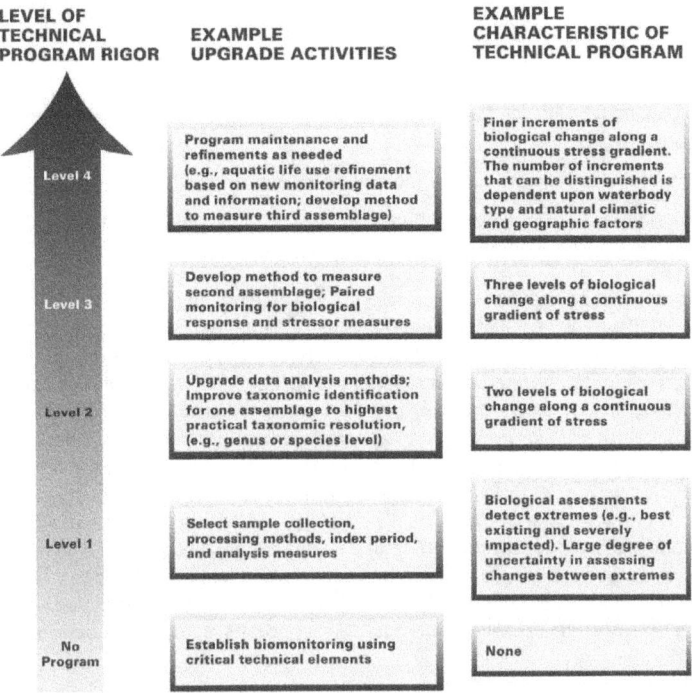

LEVEL OF TECHNICAL PROGRAM RIGOR

EXAMPLE UPGRADE ACTIVITIES

EXAMPLE CHARACTERISTIC OF TECHNICAL PROGRAM

Level 4 — Program maintenance and refinements as needed (e.g., aquatic life use refinement based on new monitoring data and information; develop method to measure third assemblage) — Finer increments of biological change along a continuous stress gradient. The number of increments that can be distinguished is dependent upon waterbody type and natural climatic and geographic factors

Level 3 — Develop method to measure second assemblage; Paired monitoring for biological response and stressor measures — Three levels of biological change along a continuous gradient of stress

Level 2 — Upgrade data analysis methods; Improve taxonomic identification for one assemblage to highest practical taxonomic resolution, (e.g., genus or species level) — Two levels of biological change along a continuous gradient of stress

Level 1 — Select sample collection, processing methods, index period, and analysis measures — Biological assessments detect extremes (e.g., best existing and severely impacted). Large degree of uncertainty in assessing changes between extremes

No Program — Establish biomonitoring using critical technical elements — None

Figure 1-2. Examples of typical upgrade activities state or tribal water quality agencies have taken to incrementally strengthen their technical programs. The example characteristics provided in column three are relevant to a biological assessment program's technical capability to distinguish incremental biological change along a gradient of increasing stress. Improved ability to discriminate biological changes supports more detailed description of designated aquatic life uses and derivation of biological criteria.

tolerant organisms (e.g., scuds, worms, snails), a pervasive algae bloom, or a fish kill. As the technical program is improved, the agency will be able to use biological assessment information with increasing confidence to more precisely define aquatic life uses, develop biological criteria, and, in conjunction with whole effluent, physical, chemical, and land use data, identify stressors and their sources.

Matching the existing level of technical rigor with the intended use of the information can provide insight on the benefit of technical development. An agency can use this understanding to guide decisions and priorities on technical development of its biological assessment program. As part of the review, agency managers and staff from the biological assessment program and other water quality programs discuss how biological assessment information is currently used to support the overall water quality management program and on program enhancements that might lead to more comprehensive and effective use of biological assessment information. On the basis of the reviews conducted beginning in 2004 (Yoder and Barbour 2009; USEPA 2010b), an agency's ability to comprehensively and effectively use biological assessment information is supported by:

- Refined aquatic life use classification to protect existing conditions and maintain improvements.

- Numeric biological criteria adopted into water quality standards (WQS).

- Coordinated biological, whole effluent toxicity (WET), chemical, and physical monitoring to support both condition assessments and causal analysis.

Program managers and staff from the monitoring and assessment programs, WQS, CWA section 305(b) report, 303(d) list, Total Maximum Daily Load (TMDL), National Pollutant Discharge Elimination System (NPDES), and nonpoint source programs jointly discuss information needs and program schedules. A water quality agency might support development of a rigorous technical biological assessment program, but if the types and quality of data, data collection, and analysis are not aligned with water quality management program information needs and implementation schedules, the information might not be most effectively used. The cross-program discussion will help reveal any gaps and inconsistencies that the agency can then address. The long-term goal is to develop a well-integrated biological assessment program that produces information with the appropriate degree of accuracy, precision, and confidence to support multiple water quality program information needs (Table 1-1). The results of these discussions do not affect the scoring of the technical elements but can inform an agency's decision on level of technical development to best support its management objectives and program priorities.

Following the review, the independent technical expert prepares a technical memorandum that describes the program's current level of rigor for the 13 technical elements and identifies the technical gaps revealed in the evaluation. In conjunction with the agency review participants, the technical expert develops recommendations to improve specific technical elements. This information helps the agency target resources more efficiently, address weaknesses, and incrementally strengthen its program to better support water quality management decisions. More information about the biological assessment review process is in chapter 3.

1.5 Benefits of a Rigorous Biological Assessment Program

As stated previously, at all levels of technical development, biological assessment information can be used to support water quality decisions. However, the degree of confidence in the use of information will increase with technical development. For example, improvements in the ability to detect changes in biological assemblages along a gradient of stress can enhance precision in describing high-quality waters and setting incremental restoration targets, as well as discriminating between intermediate levels of condition (e.g., Diamond et al. 2012). Characteristics of high level programs include improved sensitivity in the biological indices to

Table 1-1. Example discussion questions and topics on use of biological assessments to support water quality management program information needs.

Self Assessment Question	Program Implementation
Does the biological assessment program produce adequate data and information to develop biological criteria, provide detailed descriptions of designated aquatic life uses, support identification and protection of high-quality waters, and inform use attainability analysis (UAA)?	Narrative descriptions of aquatic life use classes and attendant numeric biological criteria incorporate elements of natural classification strata consistent with underlying distinction of aquatic ecotypes at appropriate spatial scale for application of the information. The biological assessment program provides data and information to define biological expectations for a specific water body or watershed and support water quality management decisions to protect existing conditions and support improvements.
How is the biological monitoring and assessment program conducted to support multiple water quality management program objectives? Does the program work with other water quality management programs to coordinate biological (including WET), chemical, and physical monitoring and assessments?	Monitoring and assessment is integrated into the overall management of surface water quality to support both determination of general condition and causal analysis. Spatial design is sufficient to detect and characterize chemical and non-chemical pollution gradients and to associate measured changes in biotic assemblages with specific or categories of stressors. Results are expressed to support multiple program uses including WQS attainment, CWA sections 305(b) reporting and 303(d) listing, CWA section 402 NPDES program, and watershed, reach, and site-specific support (i.e., investigations, watershed planning, site-specific water quality criteria development, UAA).
Is there a method developed for stressor identification and implemented as part of the water quality program? How is the information used to support multiple water quality management programs?	Empirical relationships between biological measures and chemical/physical parameters are well-developed and documented. Information is used to support statewide/regional development and refinement of water quality criteria and support stressor identification as an integral part of the assessment process. This, in turn, supports development of TMDLs.

measure incremental biological changes along a gradient of stress (Levels 3 and 4) and a more complete assessment of the community by measuring two or more assemblages (Level 4). A Level 4 program should also be able to support more expedient and robust causal analysis, because the biological assessments are coordinated with WET, chemical, and physical monitoring. Field data are linked with information on sources of stress and watershed characteristics to support source identification. Two examples of program benefits shown by states that have piloted the biological assessment review follow.

Example 1: Aquatic life use refinement. A biological assessment program with a high level of technical rigor provides for a greater degree of confidence in an agency's ability to establish biological thresholds that protect existing conditions, determine potential for improvements, and monitor to track progress and maintain improvements. For example, based on measured changes in biotic assemblages, Vermont has the technical capability to discriminate multiple increments of biological change along a gradient of stress that spans excellent to severely impacted conditions. Based on these data and information, Vermont has adopted three aquatic life use classes in its WQS (e.g., excellent, very good, good). The state has set aquatic life uses classes for its streams and rivers to maintain existing high-quality conditions. The specific use class assigned to a water body is based on its current condition, and, if degraded, its potential for improvement. Ohio has likewise adopted multiple levels of aquatic life use classes (e.g., exceptional warmwater and warmwater habitat). Additionally, Ohio has established biological expectations for agricultural drainage ditches and permanently altered streams (e.g., modified warm water habitat and limited resource waters, respectively) following a use attainability analysis (UAA) process. Ohio's use assignments undergo periodic review and upgrades based on routine, coordinated chemical, physical, and biological monitoring and assessments, including data from WET monitoring.

For both states, biological assessments conducted in conjunction with physical, whole effluent, and chemical monitoring enables them to evaluate the potential for improved conditions in their streams and rivers and consequently set appropriate and attainable goals in their WQS (e.g., designated aquatic life uses). Additionally, routine monitoring provides new data that is used to upgrade waters to a higher aquatic life use class as conditions improve (USEPA 2011c).

Example 2: Causal analysis. A finding of biological impairment does not assist management in correcting the problem unless causes of the impairment can be identified. A common use of stressor identification, or causal analysis, is in the TMDL program in situations for which a water body has been determined to have one or more impaired designated uses but the pollutants causing or contributing to the use impairments are not identified at the time. A monitoring program that collects comprehensive biological (including WET), physical, and chemical information in a coordinated manner will have the ability to examine evidence for causes of observed impairments and to develop stressor-response relationships that can inform stressor identification (e.g., Yoder and Rankin 1995b; Suter et al. 2002). For example, the Maine Department of Environmental Protection (MDEP) evaluated the condition of the Pleasant River watershed with biological indices for benthic macroinvertebrates and algae in combination with chemical and physical data and information. Located in southern Maine, the Pleasant River watershed is primarily forested with some agriculture and increasing amounts of residential development in the downstream portions of the watershed. The Pleasant River has a water quality goal of Class B—good quality conditions.

MDEP sampled algal and macroinvertebrate communities in several locations on the Pleasant River. Biological assessment results showed that the headwater reach attained Class B. Further downstream, the macroinvertebrate samples attained Class B. However, some of the downstream algal samples attained a lower level of quality comparable to Class C conditions (i.e., waters in fair condition). The river segment was also listed as impaired because it did not attain the Class B dissolved oxygen criterion. MDEP used water chemistry data, habitat evaluations, and diagnostic algal and macroinvertebrate metrics to determine that phosphorus enrichment was the probable stressor for these downstream sites. To prepare for developing a TMDL, MDEP evaluated the watershed and identified some farms and residential areas as potential sources of nutrients in the lower part of the watershed. The combination of biological assessments for multiple taxonomic groups and associated chemical, habitat, and land use information allowed MDEP to complete a thorough and more expedient evaluation of the Pleasant River watershed. As a result, MDEP has started developing a TMDL that will effectively target management actions needed to maintain biological conditions in the headwaters and to restore downstream portions of the watershed.

Use of multiple biological assemblages and coordinated biological, WET, chemical, and physical monitoring are characteristics of a Level 4 biological assessment program, and these capabilities can lead to improved confidence in estimating stress-response relationships. A relational database that enables data export and analysis via query supports this function. This level of technical development improves an agency's efficiency in identifying water quality limited waters that must be placed on a state or tribe's CWA section 303(d) list, conducting causal analysis, and assigning probable cause, or causes, of impairment. As a result, an agency should be able to more efficiently develop the appropriate management action to address a TMDL (or suitable alternative means of achieving WQS) when a pollutant has been identified as the cause of a biological impairment. A well-established, well-supported, and comprehensive monitoring program then provides the data needed to track progress and evaluate the effectiveness of the management actions taken, whether monitoring discharges and tracking the effects of permit limits or monitoring the implementation of best management practices (BMPs) for nonpoint source pollution. Paired stressor-response data might also be used to develop or refine chemical water quality criteria (Cormier et al. 2008; USEPA 2010c), and it has been used to identify benchmarks for conductivity (USEPA 2011a).

Overall, a monitoring program that integrates biological assessment, WET, chemical, and physical data is key for the most effective implementation of the biological assessment program and supports use of biological assessments to more precisely define aquatic life uses and derive numeric biological criteria. Additionally, when the monitoring schedule coincides with the cycle of WQS establishment and review, CWA section 305(b) reporting and section 303(d) listing, TMDL development, NPDES permitting, and nonpoint source program implementation, biological and other environmental data are available when needed by water quality management programs. Several states have improved cross program coordination through a rotating basin approach.

A well-established biological monitoring and assessment program will further benefit an agency's water quality program if comparable or consistent sample collection methods and data analysis protocols are developed in conjunction with the biological monitoring programs of other agencies (e.g., at local level and adjacent states, tribes; federal). This approach will support development of regionally consistent taxonomy for biological data and will help address data gaps regarding regionally appropriate, taxon-specific tolerance values and other ecological traits. Such consistent data allow for shared use of reference site data across jurisdictional boundaries. In some places there is a paucity or total lack of reference sites comparable to minimally disturbed conditions. The ability to share data and expand reference site network beyond jurisdictional boundaries might support establishing more robust reference conditions.

1.5.1 Implications for Technical Program Development

The technical capabilities of Level 1 and 2 programs are appropriate for some, but not all, water quality program uses. For example, a Level 1 program can typically differentiate water bodies in the very best and worst conditions, whereas a Level 2 program can more confidently assess good and poor conditions. Both these programs can make defensible determinations of failure

9

to fully support a water body's designated aquatic life use, but they might fail to detect initial and significant changes in biological condition caused by anthropogenic stress. Some degraded water bodies might not be accurately assessed, and, therefore, no actions are initiated to remediate and restore them. Southerland et al. (2006) estimated that up to 25 percent of impaired sites would escape detection (i.e., would pass as unimpaired, or false negatives) simply from lax reference site-selection criteria. This situation is of particular concern if a threshold is selected at the low boundary of a reference condition.

1.5.2 Benefits of a Biological Assessment Program Review

An agency can use the biological program review to determine the capabilities of its biological assessment program in a consistent, systematic manner that supports further technical development and enables midcourse review and refinement. The review will help determine if information is collected and analyzed with the accuracy and precision appropriate to address a variety of water quality management issues. The agency will be able to propose refinements to its water quality program to enable more comprehensive and efficient use of biological assessment information to support water quality management in a variety of water quality programs (e.g., NPDES permitting, TMDLs). This process and its outcomes help communicate the value of further technical development to agency management and to the public. The process, steps, and workshop materials for the biological program review are further discussed in chapters 2 and 3 of this document.

CHAPTER 2: THE TECHNICAL ELEMENTS OF A BIOLOGICAL ASSESSMENT PROGRAM

A biological assessment program's level of rigor is dependent on the quality and level of resolution of 13 technical elements (Table 2-1).

Table 2-1. Definitions of the technical elements

	Technical Element	Definition
Biological Assessment Design	Index Period	A consistent time frame for sampling the assemblage to characterize and account for temporal variability.
	Spatial Sampling Design	Representativeness of the spatial array of sampling sites to support statistically valid inference of information over larger areas (e.g., watersheds, river and stream segments, geographic region) and for supporting water quality standards (WQS) and multiple programs.
	Natural Variability	Characterizing and accounting for variation in biological assemblages in response to natural factors.
	Reference Site Selection	Abiotic factors to select sites that are least impacted, or ideally, minimally affected by anthropogenic stressors.
	Reference Conditions	Characterization of benchmark conditions among reference sites, to which test sites are compared.
Data Collection and Compilation	Taxa and Taxonomic Resolution	Type and number of assemblages assessed and resolution (e.g., family, genus, or species) to which organisms are identified.
	Sample Collection	Protocols used to collect representative samples in a water body including procedures used to collect and preserve the samples (e.g., equipment, effort).
	Sample Processing	Methods used to identify and count the organisms collected from a water body, including the specific protocols used to identify organisms and subsample, the training of personnel who count and identify the organisms, and the methods used to perform quality assurance/quality control (QA/QC) checks of the data.
	Data Management	Systems used by a monitoring program to store, access, and analyze collected data.
Analysis and Interpretation	Ecological Attributes	Measurable attributes of a biological community representative of biological integrity and that provide the basis for developing biological indices.
	Discriminatory Capacity	Capability of the biological indices to distinguish different increments, or levels, of biological condition along a gradient of increasing stress.
	Stressor Association	Relationship between measures of stressors, sources, and biological assemblage response sufficient to support causal analysis and to develop quantitative stress-response relationships.
	Professional Review	Level to which agency data, methods, and procedures are reviewed by others.

The following section describes each technical element and provides a template for assigning a level of technical rigor to each element. Section 2.2 describes how these scores are summarized to estimate an overall level of technical rigor for a biological assessment program.

2.1 The Technical Elements

2.1.1 Index Period: Characterizing and Accounting for Temporal Variability (Element 1)

(Lowest) 1.0	2.0	3.0	4.0 (Highest)
Temporal variability is not taken into account.	Sampling period established based on practices of other agencies and/or literature. Sampling outside the index is not adjusted for temporal influence.	Index period established based on a priori assumptions regarding temporal variability of biological community. Effects of the use of index period are documented. Data collected outside the index period data might be adjusted to correct for temporal influences.	Temporal variability is fully characterized and taken into account for all data. Agency information needs and index periods are coordinated so that adherence to an index period is strict.

Biological communities vary over time due to the life cycles of the targeted organisms (e.g., reproduction, recruitment, growth, emergence, and migration) and temporal variations in environmental conditions (e.g., changes in flow), so the characteristics of a biological sample can also vary depending on when that sample is collected. This temporal variability must be taken into account when interpreting biological data and assessing biological condition. Two approaches are commonly used: index periods and continuous models.

An index period is a contiguous time period used to minimize variation among biotic samples associated with systematic phonological changes in population densities and assemblage structure (Munné and Prat 2011; Kosnicki and Sites 2011). Selection of an index period can be based on a priori, existing knowledge regarding the predictable temporal changes in assemblage structure described above, when resident populations are comparatively stable (e.g., periods of growth between recruitment and emergence), and when potential exposure to anthropogenic stressors is highest (e.g., Resh and Rosenberg 1984, 1989; McElravy et al. 1989; Barbour et al. 1996; Bailey et al. 2004; Bollmohr and Schultz 2009). The index period can be further refined or based on analysis of data collected throughout the year to identify those periods in which assemblage composition is most stable. When selecting an index period, a biological assessment program also typically considers availability of sampling crew and accessibility to and safety of sampling sites.

Continuous models can also be used to characterize and account for natural temporal variations in the characteristics of biological assemblage. These statistical models estimate relationships between different biological attributes and the season or day of the year when the samples were collected (e.g., Hawkins 2006). For example, day of the year was the single most important predictor in development of an observed/expected (O/E) index in North Carolina, and the O/E model was adjusted for phonological shifts in species abundance (Hawkins 2006). The day of the year was the single most important predictor in development of the O/E index and the model adjusted for phonological shifts in species abundance. Continuous models can be applied to data collected in index periods or across multiple seasons. Indeed, approaches that combine data collected during index periods with models to account for temporal variations within index periods are often the most effective means of accounting for temporal variations. Also, one can calibrate multiple seasonal indicators and indexes, or develop an average or composite annual characterization based on multiple samples (e.g., Furse et al. 1984; Linke et al. 1999; Cao and Hawkins 2011; Pond et al. 2012).

Scoring of the index period element depends on how thoroughly a program has considered and documented the effects of different index periods on the characteristics of biological data and on decisions derived from this biological data. Example evaluation questions are:

- Is sampling carried out primarily within a defined index period?
    - If not, are the program's indices structured to account for temporal variability?
- What are the justifications for the defined index period, and has variability within the index period been quantified?
- If an alternative approach has been selected, does this approach adequately account for temporal variability?
- Are the monitoring and other water quality management programs coordinated their schedules so that data are provided when the programs need it? Does lack of coordination result in monitoring outside of the index period?

Programs that score highly on this element have documented the effects of the index period or an alternative approach to address temporal variability. Additionally, the monitoring and other water quality management programs have coordinated their schedules so that program information needs (e.g., condition assessments, permit reviews, total maximum daily load [TMDL] development) are coordinated with data delivery.

Frequently Asked Question

Question: What is the optimal time of year to select as an index period?

Answer: Selection of an index period is part of the overall design process that takes into account scientific knowledge, objectives, costs, logistics, and information desired from the monitoring program (Hughes and Peck 2008). For example, seasonal phenology influences the species composition in streams; late-instar (and hence easy to identify) stoneflies and mayflies occur in early spring, but in early summer they might be present only as very small early instars (e.g., McCord and Lambrecht 2006). Fish sampling is generally avoided in the spawning seasons of anadromous fish (Hughes and Peck 2008). Safety and logistics are also issues, as is scheduling the sequence of field, laboratory, data processing, and reporting tasks; sampling might be dangerous during the spring freshet (snowmelt), and high elevation streams might only be accessible in the summer (Hughes and Peck 2008). As depicted in Table 2-2, the index period can vary by state and assemblage group.

Table 2-2. Examples of biological assessment index periods for different state water quality agencies

	Winter			Spring			Summer			Fall		
	Dec	Jan	Feb	Mar	Apr	May	Jun	Jul	Aug	Sep	Oct	Nov
Vermont (Benthos)										▓		
Vermont (Fish)								▨	▨	▨		
New Jersey								▓				
Maryland (Benthos)				▓	▓							
Maryland (Fish)							▨	▨	▨			
Mississippi		▓										
New Mexico				▓	▓	▓	▓	▓	▓	▓	▓	
Iowa (Benthos)								▓				
Iowa (Fish)								▨	▨	▨	▨	
Arizona					▓	▓						
Idaho							▓	▓	▓			

Benthos ▓   Fish ▨

2.1.2 Spatial Sampling Design (Element 2)

(Lowest) 1.0	2.0	3.0	4.0 (Highest)
Study design consisting of isolated, single, fixed-point sites.	Low density fixed station design. Multiple sites are used for assessment of a water body or watershed condition. Spatial coverage suitable for general condition assessments. Non-random designs at coarse scale used (e.g., 4–8 digit hydrologic unit code [HUC]). Inference of site data to larger unit of assessment based on "rules of thumb" and might be supplemented by upstream/downstream assessments.	Low density random or stratified random sampling design which allows for a statistically valid inference of biological condition to a spatial unit larger than a site. The primary goal is to assess aggregate condition and trends on a statewide or regional basis.	High density (e.g., intensive) monitoring at comprehensive spatial sampling design suitable for watershed assessments (e.g., 10–12 digit HUC) and in support of multiple water quality management program needs for information (e.g., condition assessments, use refinement, use attainability analyses [UAAs], permits). As needed, the spatial sampling combines monitoring designs to optimize cost and efficiency in data collection and analysis (e.g., combination of upstream-downstream, intensive, probabilistic, and/or pollution gradient designs). Typically includes a rotating sequence of watershed units organized to provide data for management program support.

Water quality programs have multiple needs for information (e.g., status and trends, stressor identification, targeted studies, discharge monitoring). This technical element addresses how well a biological assessment program is able to (1) deploy monitoring designs that address the suite of water quality program information needs; (2) cover the pollution gradients that are relevant to the impairments that are detected; and (3) provide data relevant to the scale required for specific management program needs (e.g., stream segment, watershed, region, statewide) and that support statistically valid inferences of site data to the unit of assessment.

Study design pertains to the spatial array of sampling sites to support assessments at watershed and stream- or river-segment specific scales. It also includes the ability to provide biological assessment data and information to address multiple water quality program questions (e.g., status and trends, environmental outcomes of management actions, as well as relevant targeted studies such as discharge monitoring and TMDL implementation) at the same scale at which management is being applied. A biological assessment program will need to determine what sampling design, or combination of sampling designs, will provide the full suite of information needed to address its priority management questions (e.g., for site-specific use attainability determinations, biological criteria derivation, targeted assessments, causal analysis, statewide and regional status).

Whether single or multiple sampling designs are employed, they will need to support multiple management program support tasks. Multiple, overlapping monitoring designs can be appropriately scaled to address these specific needs when the designs are incorporated into an overall spatial network for monitoring (e.g., upstream-downstream; intensive, probabilistic, gradient design). For example, sampling upstream and downstream of a discharge is conducted to specifically quantify the effects of that discharge. A gradient design is appropriate for refinement or development of biological or other types of water quality criteria. Spatially intensive sampling can be designed for specific studies and purposes including site-specific criteria development or refinement. A probabilistic monitoring design can be tailored for condition assessments at different spatial scales (e.g., watershed, basin, ecological region, statewide). In some cases, with upfront planning, the monitoring designs can be complementary with sampling sites providing data relevant to more than one purpose.

Study designs also need to factor in adjustments for effects of natural gradients. This adjustment is typically accomplished iteratively when accounting for natural spatial variability (see technical element three) and dependent upon assessment objective (e.g., define stressor gradient, assess condition, determine cause of impairment in a stream segment). For example, in streams and rivers, the structure of aquatic assemblages changes naturally and predictably as one moves downstream from steeper, narrow, shaded, small steams to low-gradient, open-canopied, large streams (Vannote et al. 1980). Sampling sites might be located in linear juxtaposition to one another in a river or stream network. In these situations observations at nearby sites might be spatially autocorrelated and, hence, not statistically independent of one another (e.g., NAS 2002). These considerations should be addressed in the spatial sampling design and in subsequent analysis of data to accurately and precisely define the expected biological community for a water body (e.g., refined aquatic life use) and to minimize risk of making nonattainment decisions on the basis of natural changes in assemblage as one samples further downstream.

Scoring of this technical element is based on the degree to which the selected sampling sites can inform multiple water quality information needs and support decisions at different spatial scales. Example evaluation questions are:

- Is the spatial study design sufficient to represent the majority of water types in the area of interest?

- Are all pollution impacts and gradients adequately characterized?

- For condition assessments, how well can inferences be made to unsampled sites within the unit of assessment (e.g., site, stream segment, watershed, basin, statewide, ecological region)?

- For specific water bodies of concern, can valid inferences be made on differences in condition upstream and downstream of a discharge, and on changes before and after implementation of best management practices (BMPs)?

Programs that achieve high scores on this technical element have implemented an integrated sampling design, or combination of sampling designs, that provide the data and information necessary to support water quality management decisions at multiple spatial scales (e.g., specific sites, entire watersheds, basins, ecological regions, statewide).

Frequently Asked Questions

Question: What type of study design can efficiently support statewide condition assessments and 305(b) reports?
Answer: A probabilistic sampling design can be used to randomly select sampling sites from the population of water bodies so that inferences from this random subsample can be made to the entire population (Herlihy et al. 2000; Olsen and Peck 2008). A probabilistic design is the most efficient sampling design for statewide condition assessments such as the Clean Water Act (CWA) section 305(b) reports since all potential sampling locations have a known probability of being selected and inference to larger geographical area is statistically robust (e.g., Thompson 1992; Olsen et al. 1999; Olsen et al. 2009). When resources are not available to sample all basins statewide in any particular year, a rotating basin approach can be implemented.

Question: What type of study design can support assessing use designations, conducting use attainability analyses (UAAs), and providing information about multiple stressors at a watershed scale?

Answer: There are several sampling designs that could be used when appropriately designed to answer these questions, including a survey, gradient, or random designs tailored to the appropriate spatial scale. For example, a geometric and intensive watershed

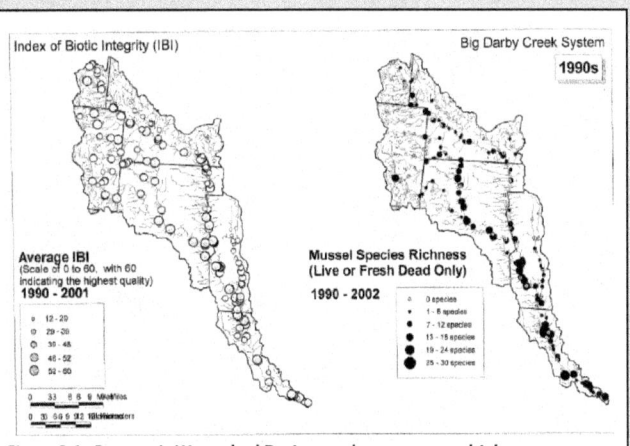

Figure 2-1. Geometric Watershed Design used to support multiple management needs in the Big Darby Creek watershed, Ohio (Ohio EPA 2004).

design was used at the 11-digit hydrologic unit code (HUC) scale in Big Darby Creek, Ohio, and, when considering serial autocorrelation between adjacent sites, is nearly equivalent to a census of the stream reaches of the watershed (Figure 2-1). The data were used to determine if the current aquatic life use of stream and river segments was appropriate and attainable and then to determine the status of each site. The data were also used to delineate impairments for reporting (e.g., CWA section 305[b]/303[d]), and causes and sources were determined to support specific water quality management actions (i.e., TMDLs, National Pollutant Discharge Elimination System [NPDES] permits, stormwater permitting, 401 certifications) and support watershed planning (i.e., section 319 planning and implementation). Ohio conducts four to five of these assessments annually with a rotating basin approach, and, in the aggregate, each contributes to a statewide inventory of streams and rivers and is part of a database that supports many program maintenance and developmental needs. These data are aggregated upwards to produce regional and statewide assessments for meeting CWA 305b reporting and internal program goal tracking (e.g., the Ohio 2020 goals).

Question: What are the benefits of combining probabilistic design surveys with intensive surveys designed to answer multiple water quality management questions?

Answer: Developing the technical capacity to conduct different types of survey designs enhances the breadth and depth of the monitoring program's ability to answer multiple water quality management questions and to more efficiently leverage resources. For example, in 2008, New York State Department of Environmental Conservation's (NYSDEC's) Stream Biomonitoring Unit merged a random probabilistic design survey with its legacy statewide basin studies. This *hybrid* survey design allows it to fit the needs of two primary objectives of its program: surveying targeted of-interest sites, and creating an unbiased random data set (Figure 2-2). Targeted sites include those that allow for the characterization of regional reference conditions, long-term temporal trend monitoring, assessment of unassessed waters, and the monitoring of sites that are of department, regional, and/or public interest. The random data set gives the ability to project aquatic life use attainment in an un-biased, statistically sound manner across the entire state, and provides uniform comparability between basin data sets and other national data sets. Targeted sites make up approximately 60 percent of the total number of sites sampled each year while random sites compose 40 percent.

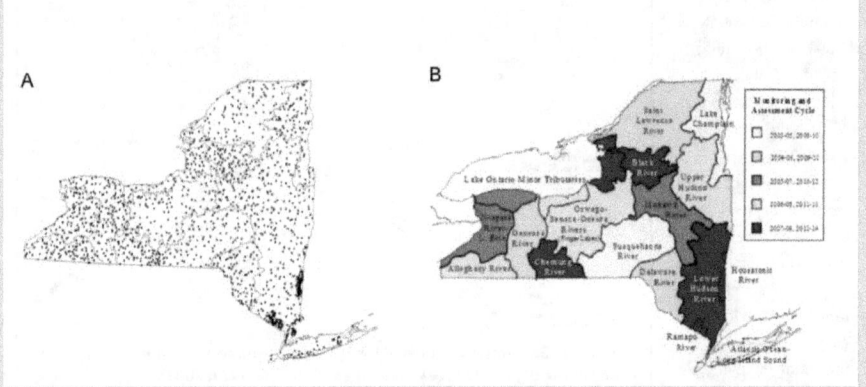

Figure 2-2. New York has integrated a probabilistic spatial survey design (A) into its routine rotating integrated basin studies program (B) (Source: NYSDEC 2009).

2.1.3 Natural Variability: Characterizing and Accounting for Spatial Variability (Element 3)

(Lowest) 1.0	2.0	3.0	4.0 (Highest)
No or minimal partitioning of natural variability in aquatic ecosystems. Does not incorporate differences in watershed characteristics such as size, gradient, temperature, elevation, etc.	Classification scheme is based on assumed, first-order classes. These include strata such as fishery-based cold or warmwater classes. There is no formal consideration of regional strata such as bioregions or aggregated ecoregions. Intra-regional strata such as watershed size, gradient, elevation, temperature are not addressed. Usually applied uniformly on a statewide basis.	A fully partitioned and stratified classification scheme or modeling approach is employed. Classes and/or continuous models are defined to take critical details of spatial variability into account. Inter-regional landscape features and phenomena are appropriately sequenced with intra-regional strata. Subcategories of lotic ecotypes are defined (e.g., includes the full strata of lotic water body types). Characterization of spatial variability is confined within jurisdictional boundaries.	Scheme to fully account for natural variation is periodically refined and updated as new data and methods become available. Classes, continuous models, or both, are examined to identify the most appropriate scheme for monitoring and assessment, regulatory support, and cost-effectiveness. Developed at scales that transcend jurisdictional boundaries when necessary to strengthen inter-regional classification outcomes; recognizes the full zoogeographical aspects of biological assemblages.

Biological assemblage structure varies spatially among different sites, often associated with variations in abiotic environmental conditions (Theinemann 1954; Hynes 1970; Poff 1997). Both local (e.g., water temperature, flow, and alkalinity) and regional environmental conditions (e.g., basin topography, climate) strongly influence assemblage structure, and when interpreting biological data and assessing condition, natural variations in assemblage structure must be characterized and taken into account to ensure that changes in assemblage structure can be confidently attributed to anthropogenic rather than natural factors.

Well-developed schemes to account for natural variation use a combination of large-scale physical characteristics (e.g., watershed drainage size, elevation, geographic location) and local site characteristics (e.g., temperature, alkalinity, substrate) (Moss et al. 1987; Reynoldson et al. 1997; Bailey et al. 1998; Marchant et al. 1999; Joy and Death 2002; Hawkins et al. 2000a; Oberdorff et al. 2002). The principal approaches used are classification (or typology), continuous models, and combinations of discrete and continuous models.

Classification schemes define classes of water bodies such that sites in each class are assumed to be similar with one another in terms of naturally varying abiotic factors. Then, biological assemblages observed at sites in each class are examined to determine if they are more similar to one another than among classes. These classes can be defined *a priori* based on an ecological understanding of natural factors that structure biological assemblages (Omernik 1987; Rabeni and Doisy 2011) to help design sampling strategies that represent all water body types in a study area. Classification schemes can also include classes of water bodies that pertain to inherent environmental requirements (e.g., warm and cold water, strata), differences in discrete lotic strata (headwaters to large rivers), and continuous changes in assemblage structure across natural environmental gradients (e.g., Moss et al. 1987). Classes can also be specified *a posteriori* by statistically examining how assemblage structure varies across different environmental gradients and defining discrete classes based on the results of these analyses (Gerritsen et al. 2000). In either case, the biological condition at a particular site is assessed by comparing to reference conditions in the class to which the site belongs.

Natural variations in assemblage structure can also be taken into account using models that represent changes in structure over continuous environmental gradients (Growns 2009; Hawkins and Vinson 2011; van Sickle and Hughes 2000). These models are based on statistical analyses that can be used to infer changes in assemblage structure due to different environmental variables (Clarke et al. 1996; Bailey et al. 1998; Marchant et al. 1999; Hawkins et al. 2000b; Simpson and Norris 2000; Joy and Death 2002). When a model is used to assess a site, a site-specific prediction of biological characteristics is calculated, and the observed characteristics assessed relative to this prediction. This information can also be used to supplement or refine discrete classification approaches.

A comprehensive classification and/or modeling scheme is dependent on the spatial density of the monitoring program. Sufficient spatial coverage is needed to test or verify a proposed classification and/or modeling scheme (see Technical Element 2).

Scoring of Technical Element 3 is based on the degree to which the scheme accounts for observed natural variability in biological assemblage structure. Example evaluation questions are:

- Does classification or modeling the effects of natural gradients sufficiently reduce natural variability relative to anthropogenic variability?
- Does the classification scheme and/or modeling process sufficiently include all the common regional and watershed strata in the study area?
- Is the approach sufficient to support the precision and accuracy needed in estimates of biological index values?
- Does the classification and/or model take into account information and considerations from beyond a state or tribe's jurisdictional boundaries?

Programs that score highly in this technical element have demonstrated that their scheme to describe natural variability (whether classification and/or continuous models) accounts for the major sources of natural variability in the study area, and that the majority of the remaining variability in biological characteristics can be attributed to human activities.

Frequently Asked Questions

Question: What is meant by an ecoregional classification for biological assessment?
Answer: Partitioning the water bodies of an agency by natural variability in the biota results in a classification that can improve assessment of ecological condition. As an example, natural classification in Mississippi resulted in five bioregions (not counting the delta region in gray) as a basis for biological assessment (Figure 2-3). Bioregions are geographically distinct regions of water bodies that roughly correspond to ecoregions or aggregations of ecoregions.

Question: How would a multivariate cluster analysis serve as a form of classification?
Answer: Clustering the biological data from reference sites reveals the inherent natural variability among of sites. Clusters can be selected that represent classes for assessment membership.

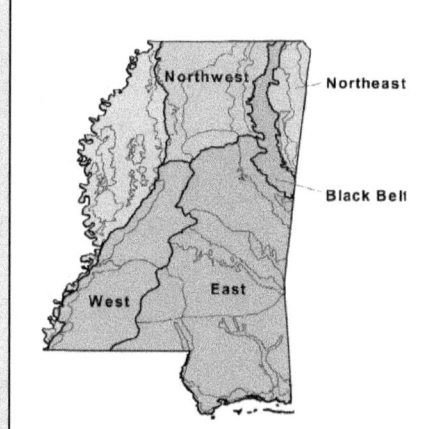

Figure 2-3. Example of bioregions as established for the Mississippi.

2.1.4 Reference Site Selection (Element 4)

(Lowest) 1.0	2.0	3.0	4.0 (Highest)
Informal best professional judgment (BPJ) used in selection of control sites. No screens are used. Limited, if any, documentation and supporting rationale.	Based on "best biology" (i.e., BPJ on what the best biology is in the best water body). Minimal non-biological data used. Minimal documentation.	Selection based on narrative descriptions of non-biological characteristics. Combines BPJ with narrative description of land use and site characteristics. Might use chemical and physical data thresholds as primary filters.	Based on quantitative descriptions of non-biological characteristics with primary reliance on abiotic data on landscape conditions and land use. Chemical and physical data might be used as secondary filters or in a hybrid approach for severely altered landscapes. Independent data set used for validation.

Reference site selection is the basis for developing benchmarks against which a biological monitoring program can assess the biological condition of test sites (e.g., Hughes et al. 1986; Barbour et al. 1996; Bailey et al. 2004; Stoddard et al. 2006; Hawkins et al. 2010). Reference site selection is primarily based on abiotic factors that define sites that are "least stressed," or ideally, "minimally stressed" by anthropogenic stressors and include knowledge of whether invasive species are present (e.g., Hughes et al. 1986; Karr and Chu 1999; Bailey et al. 2004). Abiotic characteristics and attributes should be the principal screens for selecting candidate reference sites because such screens avoid circularity that is inherent in including ambient biological characteristics to define reference sites for assessing biological condition.

Factors to be considered in selecting reference sites include human population density and distribution, proximity to the influence of discharges, proximity to physical modifications of stream and river channels, road density, and the proportion of mining, logging, agriculture, urbanization, grazing, or other land uses. Candidate reference sites are evaluated with respect to these factors to determine the degree of human modification that has occurred. Sites that are minimally disturbed by potential stressor(s) are considered to be in reference condition (Bailey et al. 2004; Stoddard et al. 2006). Ideally, sites are eliminated if they have undergone direct human modification, especially to riparian zones and instream habitat (Bryce et al. 1999). However, in some pervasively altered regions or altered systems, "least disturbed" sites that represent the best available conditions have been used (e.g., Angradi et al. 2009).

Examples of evaluation questions are:

- Do factors for reference site selection emphasize abiotic measures of anthropogenic activity?
- Are procedures for selection of sites well documented? Do those procedures include consideration of watershed development, near stream development, and riparian condition?
- Are chemical, physical, and whole effluent toxicity (WET) sampling data used to validate either the absence of anthropogenic disturbance or the level of allowed disturbance?

Programs that score highly in this technical element use several layers of abiotic filters to identify reference sites for their study area, primarily based on landscape data from the surrounding catchment and other information that characterizes the level of disturbance. Independent data sets are used to validate reference site selection.

Frequently Asked Questions

Question: How do factors for reference site selection influence calibration of a biological index or indicator and setting a threshold for biological criteria or for CWA section 303(d) listing decisions?

Answer: Biological criteria are typically derived from a reference site database (USEPA 1990, 1998, 2001). The reference site approach is typically also a basis for biological listing methodologies and for U.S. Environmental Protection Agency's (EPA's) national surveys of stream condition (Herlihy et al. 2008). The factors for reference site selection help define the quality of the reference condition (e.g., undisturbed, minimally or moderately disturbed, least disturbed) (Stoddard et al. 2006). Herlihy et al. (2008) examined the effects of different quality of reference sites from the large database of the U.S. Wadeable Streams Assessment (WSA). Poorer quality reference sites (equivalent to relaxing the factors for reference site selection to accept more sites) resulted in assessments in which more test sites were similar to reference than assessments done with reference sites selected based on more stringent site selection factors. In other words, when the reference sites are influenced by human disturbance, an agency might lose its ability to accurately define the desired biological condition and to differentiate biologically degraded sites from reference. The quality of the reference sites as defined by the factors for reference site selection can inform selection of a biological threshold. The percentile selection should be based on the degree to which human activities influence the study area. For example, in the WSA, the threshold for a specific ecological region was adjusted from 10 to 25 percent of the reference site distribution to account for the presence of pervasive human disturbance at reference sites (Herlihy et al. 2008).

Question: What if the pool of reference sites has to include sites with substantial disturbance even though the sites are least-disturbed in the context of the region? For example, in the Midwest, row crops and grain farming are the primary land use, and virtually no unaffected water bodies exist.

Answer: Regions with extensively altered landscapes might require a model to extrapolate current conditions to a reasonable reference. For example, a PCA-based regression model was used to project "true" reference in regions where all reference sites are highly altered (Herlihy et al. 2008). Kilgour and Stanfield (2006) developed regressions between biotic condition and percent impervious cover, and extrapolated biotic condition for very low impervious cover scenarios. In a slightly different approach when naturally occurring conditions can be estimated, Chessman and Royal (2004) used species responses to temperature, flow regime, and riverbed composition to predict the species composition of different rivers with given combinations of naturally occurring temperature, flow, and bed composition. In some cases, an agency might manage to the least disturbed condition and set incremental restoration targets that support improvements as technology and BMPs are applied. If appropriate, the expectations for an adjacent ecological region could be used to establish reference. For example, Ohio concluded that least affected reference sites did not exist in the Lake Huron/Lake Erie Plain (HELP) ecological region and used the biological expectations for a neighboring ecological region to determine a biological threshold. The key step is to recognize when minimally altered conditions do not exist, and then derive a reasonable alternative for deriving a protective biological criteria.

2.1.5 Reference Conditions (Element 5)

(Lowest) 1.0	2.0	3.0	4.0 (Highest)
No reference condition has been developed. Biological data are assessed using BPJ or based on the presence of targeted or iconic taxa.	Reference condition based on biology of an estimated 'best' site or water body. Single reference sites are used to assess biological data collected throughout a watershed. A site-specific control or paired watershed approach might be used.	Reference condition is based on a regional aggregate of reference site information. Data representing most of the major natural environmental gradients but limited in number and/or spatial density. Overall number and coverage of reference sites insufficient to support statistical evaluation of the biological condition at test sites.	Reference condition is based on data from many reference sites that span all major natural environmental gradients in the study area. Reference condition can be estimated for individual sites by modeling biota-environmental relationships. The number of reference sites is sufficient to support statistical evaluation of biological condition at test sites. Reference sites are resampled periodically. In highly altered regions or water body types, alternative methods are used to develop reference condition.

A primary goal for a biological assessment program is to estimate the expected biological condition (reference condition) for individual sites as accurately and precisely as possible. The reference condition serves as the benchmark for judging condition of the site and as basis for derivation of biological criteria. This technical element considers the number of reference sites that are available and the degree to which those reference sites account for natural environmental gradients (e.g., elevation, water body size) (Figure 2-4). This element also considers whether the number of reference sites is sufficient to support appropriate use designation and the derivation of numeric biological criteria. It is important to consider how well the reference site network is re-monitored and reevaluated. Reference condition should also be tracked by the periodic resampling of reference sites and as an integral function of the overall monitoring program.

Using a representative network of reference sites ensures that the assessment of a test site is based on a comparison with its most appropriate benchmark. Accordingly, development of meaningful reference conditionsalso requires an adequate spatial coverage to obtain a sufficient sample of reference sites.When sufficient reference site data are not available, assessments might not be possible or might be conducted with more uncertainty. In regions where all water bodies are severely altered, alternative methods might be used, including historical data, models, or hindcasting (e.g., Dodds and Oakes 2004; Kilgour and Stanfield 2006; Angradi et al. 2009).

Scoring of this technical element is based on the degree to which a sufficient number, or network, of reference sites are available to establish reference condition. Example evaluation questions are:

- Is the pool of reference sites sufficient to characterize the natural gradients in the study area (e.g., basin, ecological region, statewide)?

- Is the number of reference sites sufficient to support the use designation and derivation of biological criteria?

- Are reference sites systematically resampled to track changes in reference condition over time?

- In regions or water bodies with no adequate reference sites, are alternative methods used effectively (e.g., historical data, modeling)?

High level programs should demonstrate that the network of reference sites fully represents all the major natural environmental gradients in the study area and that the number of reference sites is sufficient to support both appropriate use designation and derivation of attendant biological criteria. Figure 2-4 provides an example approach for assessing the representativeness of reference sites.

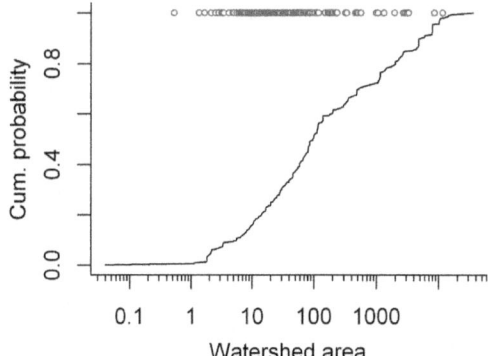

Figure 2-4. Example approach for assessing representativeness of reference sites. The solid line shows the cumulative distribution function of watershed areas for different streams in the assessed population, and the open circles show the watershed areas of the available reference sites. In this example, presence of reference sites for a watershed area is given by the density of the open circles. The majority of the watershed areas are well-represented by reference sites, because there is a high density of open circles above steep portions of the solid line; except for the largest streams (> 1,000 km^2). (USEPA 2006)

Frequently Asked Questions

Question: How does the number of reference sites (N) affect characterization of biological characteristics at a regional scale?

Answer: The number of reference sites affects both the ability to account for spatial variability (see Technical Element 3) and the precision with which thresholds can be specified. As discussed in Technical Element 3, many natural abiotic environmental factors can influence assemblage structure, and the number of reference sites directly affects the number of these factors that can be taken into account. For example, macroinvertebrate assemblage structure might vary primarily with changes in stream size (or catchment area) and, secondarily, with changes in alkalinity. Linear regression models generally require at least 10 sites per explanatory variable to accurately estimate a relationship, so at least 20 reference sites are required to model changes in assemblage structure with respect to both stream size and alkalinity. Additional reference sites that span other natural gradients would provide increased capabilities to more precisely specify natural expectations for different types of streams in the study area.

Once spatial variability is taken into account, distribution of expected index values derived from reference sites must be quantified so that index values at test sites can be evaluated. More specifically, to assess condition, one must test whether index values at a test site are within the range of index values observed in reference sites. Increased numbers of reference sites allows one to more precisely estimate the reference distribution, and therefore, more confidently assess test sites.

Question: How does the number of reference sites (N) affect the derivation of numerical biological criteria?

Answer: Determining the appropriate number of reference sites for deriving biological criteria is usually most applicable on a regional basis because of differences in reference site heterogeneity both within and between regions. In a more heterogeneous region, where natural conditions are more variable among streams, either (1) a larger reference sites pool will be necessary to accurately derive a biological criteria threshold, or (2) further partitioning of the natural variability through classification analysis might be needed. As illustrated in Figure 2-5, the variability in reference quality is reduced as the number of reference sites increases to estimate the biological criteria threshold.

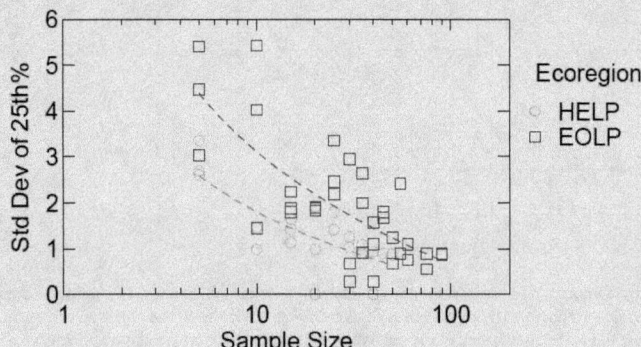

Figure 2-5. Standard deviations of 25th percentile fish assemblage Index of Biotic Integrity (IBI) scores estimated by randomly drawing reference sites at a given sample size (x-axis) five times for wading sites in the Lake Huron/Lake Erie Plain (HELP) and Erie Ontario Lake Plain (EOLP) ecoregions of Ohio (modified from Yoder and Rankin 1995a).

2.1.6 Taxa and Taxonomic Resolution (Element 6)

(Lowest) 1.0	2.0	3.0	4.0 (Highest)
One taxonomic assemblage (e.g., benthic macroinvertebrates, fish, algae, aquatic macrophytes). Very coarse taxonomic resolution (e.g., order/family). Expertise: amateur naturalist or stream watcher. Validation: none. QA/QC: none.	One taxonomic assemblage. Low taxonomic resolution (e.g., family). Expertise: novice or apprentice biologist. Validation: family level certification for macroinvertebrates. No certification available for fish or algae. QA/QC: mostly for taxonomic confirmation of voucher collections. Some sorting QA/QC implemented.	One taxonomic assemblage. Fine taxonomic resolution: genus/species for benthic macroinvertebrates and algae, species for fish. Expertise: trained taxonomist. Validation: genus-level certification or equivalent for benthic macroinvertebrates. Expert fish taxonomist or equivalent. Formal courses or training in algal taxonomy. QA/QC: addresses measuring bias, precision, and accuracy in all phases of sample processing through identification (e.g., outside validation of identification); voucher collection maintained.	Same as Level 3 except that two or more taxonomic assemblages are assessed. Rationale for selection of taxonomic groups should be well documented.

This taxonomic resolution technical element addresses the resolution to which organisms are taxonomically identified (order, family, genus, or species) and, for the highest level programs, how many different assemblages are included. Four assemblages have been primarily used in freshwater biological assessment and in making aquatic life use attainment decisions: benthic macroinvertebrates, fish, algae, and aquatic macrophytes. Methods for measuring amphibian assemblages (e.g., early life stages of salamanders) are also being developed (Moyle and Randall 1998; Whittier et al. 2007a, 2007b) for certain water body types such as primary headwater streams (Ohio EPA 2012). Each assemblage has different habitat ranges and preferences and might be susceptible to anthropogenic stressors in different manners and degrees.

As more assemblages are assessed, one can more confidently infer the condition of the entire biological community (e.g., Carlisle et al. 2008). Hence, collecting and assessing different assemblages provides a more complete assessment of the condition of aquatic life in a water body. For example, assemblages that represent more than one trophic level (primary producers, consumers, predators) might increase the ability to both assess the overall condition of the aquatic community and measure responses to multiple stressors that might affect the community. Additionally, some detectable changes in assemblages, or members of an assemblage, might provide a measure of initial stress and provide information helpful to protection of high-quality waters (e.g., Petty et al. 2010; Brooks et al. 2011; Danielson et al. 2012).

Collected organisms must be identified taxonomically before one can infer biological condition from a sample of these organisms, and the resolution of these identifications (e.g., order, family, genus, species) can influence inferences regarding the degree of biological alteration (e.g., Lenat and Resh 2001; Waite et al. 2004; Feio et al. 2006; Hawkins 2006; Pond et al. 2008; Cao and Hawkins 2011). In some cases, a finer level of taxonomic resolution allows one to better assess the sensitivity of the collected organisms to different types of stress. For example, the temperature requirements of mayflies in a certain family might vary substantially, so identifying taxa to genus or species when possible within this family might allow one to better understand the impacts of altered temperature on a water body (Vannote and Sweeney 1980). Conversely, in some regions, the number of different genera in each family might be comparatively low, so identification to family yields nearly as much information as identification to species or genus (Hawkins and Norris 2000b). In other regions, taxonomic resolution can be limited by existing taxonomic information on native fauna (e.g., Buss and Vitorino 2010). Taxonomic identification requires substantial training and practice, and quality assurance/quality control (QA/QC) of the identifications is critical for maintaining consistent standards of identification (e.g., Stribling et al. 2008).

Scoring of this technical element is based primarily on the resolution of the taxonomic identifications and on the level of QC and the number of assemblages that are routinely collected. Example evaluation questions are:

- What level of resolution is used for taxonomy and related biological attributes?
- How many assemblages are monitored?
- What training and certifications are required for persons identifying organisms?
- What are the enumeration and identification QA/QC procedures?

To score highly in this element, at least two assemblages should be used to more completely assess the condition of the entire aquatic community, and organisms should be identified to the finest practicable level of resolution. For example, for benthic macroinvertebrates this includes genus and/or species for key groups, and for fish it would include species resolution in accordance with the American Fisheries Society nomenclature (Nelson et al. 2004). Furthermore, staff who identify collected organisms should be formally trained and certified.

Frequently Asked Questions

Question: What is the best taxonomic level of identification?
Answer: The best level of taxonomic identification will vary depending on purpose of assessment and other considerations, such as the number of genera within each family in a region (Hawkins and Norris 2000b). Typically, species level is more responsive to impacts from stressors, but coarser level taxonomy can produce more precise indices (Hawkins 2006). The current ability to accurately and precisely achieve species level identification varies with the assemblage. Fish, diatoms, and macrophytes can usually be identified to species, whereas macroinvertebrates can usually be identified to genus. Lower levels of identification can improve one's ability to estimate stress-response relationships but only if that lower level of identification is not associated with a substantial increase in the uncertainty of the identifications (Stribling et al. 2008; Buss and Vitorino 2010).

Question: What is the best assemblage to assess biological condition?
Answer: Assemblages comprise different numbers and kinds of species that, in turn, differ in their sensitivities to stressors and also their occurrence and sensitivity by the water body type. The type of water body being assessed and its location (i.e., position in the landscape or river continuum) can influence the selection of assemblages to sample.

For example, small primary headwater streams (<1–10 km^2 catchment) typically have low fish species diversity, and development of fish indices can be challenging (McCormick et al. 2001; Hitt and Angermeier 2011). As such, assessing amphibian assemblage in these stream types is an alternative (e.g., Fausch et al. 1984; Moyle and Randall 1998; Whittier et al. 2007a, 2007b; Ohio EPA 2012). For wetlands, emergent macrophytes are the dominant macrobiota and are typically used for assessing wetlands (e.g., Fennessy et al. 2007), but they have also been used in rivers (Moore et al. 2012). Assemblages might also vary along the length of a waterway. For example, preferred assemblages for the Upper Mississippi River include fish, macroinvertebrates, and submerged aquatic macrophytes in the impounded portions but fish and macroinvertebrates in the open river reaches (Yoder et al. 2011).

Question: Level 4 requires 2 or more assemblages. What could the mix of assemblages include?
Answer: The mix of assemblages should be complementary rather than redundant in terms of their ecological, ecophysiological, and ecotoxicological properties (i.e., not represent the same trophic level or have the same habitat requirements). Assemblages vary in importance across water body types and respond differently to given stressors. They also respond to different intensities of the same stressor which, in turn, affects assessments of condition (e.g., Carlisle et al. 2008; Smucker and Vis 2009). For example, one approach might be to strike a balance among trophic levels: one or more animal assemblage (e.g., benthic macroinvertebrates, fish, zooplankton, benthic infauna [in estuaries]) and one plant assemblage (e.g., emergent macrophytes, floating/submerged macrophytes, periphyton, phytoplankton).

Question: Why are two or more assemblages recommended for a Level 4 program?
Answer: Measuring the response of two or more biological assemblages along a gradient of stress provides increased confidence in the program's capability to detect effects of stressors on aquatic life. There are multiple pathways in which stressors might affect the biota, and a more comprehensive measure of the biotic community provides greater confidence that these effects will be detected.
Examples of the responses of different assemblages to stressors include:

- Certain species of benthic macroinvertebrates have demonstrated consistent and measurable responses to metal toxicity. Clements et al. (2000) used cumulative criterion units to quantify metals concentrations in 95 sites in the Southern Rocky Mountain ecoregion, and they observed changes in the benthic macroinvertebrate assemblage to different levels of metals. The authors showed that highly contaminated sites had significantly lower densities of scrapers and predators and also lower in abundance and species richness of mayflies. Highly contaminated sites also had decreased abundance of mayflies, caddisflies, and stoneflies (i.e., ephemeroptera, plecoptera, trichoptera [EPT] taxa).

- A shift in species composition can signal changes in water quality. When associated with changes in levels of individual or categories of stressors, this information can be used to support identification of probable causes of biological impairment (e.g., Carlisle et al. 2008). For instance, a shift in benthic groups from those that filter the water for food to those that graze the sediments have been correlated with increase in suspended sediment load in a stream or river in absence of other stressors (Kaller and Hartman 2004). Carlisle et al. (2008) found that fish and macroinvertebrates in Appalachian streams were most sensitive to agriculture and urban land uses, while diatoms were most sensitive to chemical changes associated with mining.

- An initial increase in water column algae and shift in species composition can be an indicator of early nutrient enrichment (McCormick and Cairns 1994). Benthic diatoms have long been used as indicators of chemical water quality (e.g., Patrick 1949), and recent developments include quantitative models that infer water quality conditions from the observed diatom assemblage (e.g., Pan et al. 1996; Kelly 1998; Potapova and Charles 2003; Ponader et al. 2008; Danielson et al. 2011).

- The presence of lesions and tumors on fish can be caused by pulp and paper mill discharges (Flinders et al. 2009), pharmaceuticals (Kang et al. 2002; Lovy et al. 2007), and other types of chemicals or industrial/municipal discharges (Yoder and Rankin 1995b; Yoder and DeShon 2003). Dyer and Wang (2002) examined upstream and downstream data from 221 wastewater treatment plants in Ohio and observed impairments in fish communities downstream of large treatment plants.

- Multiple assemblages were evaluated in 268 Appalachian streams, and both fish and macroinvertebrate indices were responsive to urban and agriculturally influenced streams. Diatom assemblages were responsive to mining influence (Carlisle et al. 2008).

Question: How do we get taxonomic certification?

Answer: For some assemblages (algae, fish), professional certification of an individual's ability to accurately and precisely identify taxa is not available. However, because the accurate and precise identification of aquatic organisms is the foundation for biological assessment and monitoring programs for lakes, streams, rivers, and wetlands, certification programs are being developed. For macroinvertebrates, The Society for Freshwater Science recognized this issue a decade ago and has implemented a certification program for those professionals who identify macroinvertebrate assemblages for use in assessing aquatic habitats in North America. This program was designed to certify that trained and skilled persons are providing credible and reliable aquatic macroinvertebrate identifications at the genus and/or family level. The certification program tests a candidate's knowledge and skills in aquatic macroinvertebrate taxonomy and provides the successful applicant with a certificate of proficiency.[2]

Selected states might also offer certifications that address taxonomic and other biological assessment skills and qualification. For example, Ohio offers certification as a Qualified Data Collector under the Ohio Credible Data Law. Three levels are offered: Levels 1, 2, and 3. Level 3 is required for acceptance of data by Ohio Environmental Protection Agency (Ohio EPA) for CWA section 303(d) listing and use designation assignments under the Ohio Water Quality Standards (WQS). The certification is obtained by completing a required training class and then completing performance-based testing for fish (including habitat assessment) or macroinvertebrate assemblage assessment. Certification is also available for the Primary Headwater Habitat assessment methodology and for chemical/physical sampling. Additionally, California has developed a process to document the quality of the taxonomic identifications directly. Re-identification of a percentage (typically 10 percent) of taxonomic data by a QC laboratory is routinely required of most projects in California. Summaries of discrepancies are stored with the original data, providing users of the final data set with direct information about the quality of the original data, much as QA batch data provides information about chemistry analyses. In effect, California audits the data instead of the data providers. California also requires that taxonomists who provide data for the state be active members of the Southwest Association of Freshwater Invertebrate Taxonomists and follow its standard taxonomic effort protocols and reporting standards.[3]

Question: What is DNA barcoding, and is there potential for future application in biological assessments?

Answer: DNA barcoding is a technique by which organisms (fish, macroinvertebrates, macrophytes, algae) can be catalogued into species based on the nucleotide sequence of one or more gene (e.g., the mitochondrial c oxidase I gene for fish and macroinvertebrates). A recent approach to characterize the composition, and possibly the health of communities, is integrating DNA barcoding with metagenomics. Metagenomics refers to the technique developed to sequence all genetic material present in an environmental sample (soil or water). Moreover, next generation sequencing technology is allowing for the DNA of all species in a sample to be isolated and sequenced at once (i.e., resulting in a metagenome). Once a metagenome is obtained, sequencing of a specific gene region (barcoding) allows one to distinguish the species composition of organisms at a specific location. However, this approach cannot currently provide information regarding the relative abundance of the species present in the collections, which is an important factor in using species level data for water quality monitoring. One long-term goal of the DNA barcode approach is to link biodiversity with existing knowledge of species susceptibilities and tolerances to environmental stressors so that one can describe and evaluate the condition of a community given its biological signature.

---

[2] http://www.nabstcp.com/
[3] http://swamp.mpsl.mlml.calstate.edu/resources-and-downloads/standard-operating-procedures

2.1.7 Sample Collection (Element 7)

(Lowest) 1.0	2.0	3.0	4.0 (Highest)
Approach is cursory and relies on operator skill and BPJ. Training limited to that which is conducted annually for non-biologists who compose the majority of the sampling crew. Methods are not systematically documented as standard operating procedures (SOPs).	Textbook methods are used without considering the applicability of the methods to the study area. SOPs to specify methods but methods are neither well documented nor evaluated for producing comparable data across agencies. A cursory QA/QC document might be in place. Training consists of short courses (1–2 days) and is provided for new staff and periodically for all staff.	Methods are evaluated for applicability to study area and refined (if needed). Detailed and well documented SOPs are updated periodically and supported by in-house testing and development. A formal QA/QC program is in place with field replication requirements. Rigorous training required for all professional staff.	Same as Level 3, but methods cover multiple assemblages. A field audit of sampling crews is performed annually to ensure that protocols and proper sample handling/documentation are followed.

The sample collection technical element consists of standard operating procedures (SOPs) used to collect and preserve biological samples and take field measurements. Standardized and well-tested field methods minimize the variability in biological samples associated with differences in sampling procedures. A robust QA/QC system provides assurance that SOPs are followed. Numerous studies have demonstrated that the field methods used can have strong effects on the characteristics of the collected organisms. For example, samples collected in slow water, depositional areas provide a different set of taxa compared with samples collected in riffles (Parsons and Norris 1996). As such, for benthic macroinvertebrates, sampling protocols should specify how different habitats in a stream reach are selected for sampling (Gerth and Herlihy 2006; Rehn et al. 2007). Similarly, greater sampling effort (e.g., more time spent collecting) results in larger numbers of individuals and taxa. Use of different sampling equipment (e.g., kicknets vs. Surber samplers) alter the characteristics of the collected assemblage (e.g., Stark 1993; Cao et al. 2007; Cao and Hawkins 2011).

Scores for this technical element are based on the extent of standardization and evaluation of field sampling methods and the completeness of the QA/QC system. Example evaluation questions are:

- Are standardized methods used to select sampling locations (e.g., single or multiple habitats, transects) within a selected site and to collect and preserve samples?
- How is QA/QC incorporated in sample collection?

Biological assessment programs that score highly for this technical element have developed well-defined and rigorous SOPs that specify details of the collection (e.g., where samples are collected, what sampling equipment should be used, when samples should be collected, how samples should be preserved). The QA/QC system should provide for regular audits of field crewand replication of samples at a certain proportion of sites, assign responsibility, define personnel qualifications, establish protocols, define preventative and corrective action, provide information tracking, and ensure that study objectives are met (USEPA 1995; Stribling et al. 2008). Voucher specimens are retained to verify the accuracy of taxonomic identifications.

Frequently Asked Questions

Question: How does sample collection influence the rigor of a biological assessment?

Answer: Sample collection is the genesis of biological assessment data; therefore, how it is designed and executed influences the ability of a biological assessment to adequately and accurately describe biological quality. However, biological assessment sample collection should be sufficiently cost-effective so as to produce a sample with 2–3 hours' effort in the field.

Question: How do I know which method is best for my biological indicator (Figure 2-6)?

Answer: Methods should have a well-developed SOP, and all field personnel should be trained by qualified professionals. The SOP should minimize the decisions that need to be made in the field, and the training should provide guidance for how to handle unusual situations. If well-developed SOPs and training are done by qualified professionals with appropriate checks and/or audits in place, the actual sampling could be done by more junior personnel under the direction of senior level staff. This type of apprenticeship or mentoring is important for maintaining consistency in sample collection and minimizing variability due to who is doing the sampling at any one location and/or time.

Figure 2-6. Stream sampling methods.

2.1.8 Sample Processing (Element 8)

(Lowest) 1.0	2.0	3.0	4.0 (Highest)
Organisms are sorted, identified, and counted in the field using dichotomous keys.	Organisms are sorted, identified, and counted primarily in the field by trained staff. Adequate QA/QC is not possible. For fish, cursory examination of presence and absence only. Agency SOPs not developed or published.	All samples (except for fish) are processed in the laboratory. A formal QA/QC program is in place. Rigorous training is provided. Voucher organisms are retained for ID verification. SOPs are published and available to others.	Same as Level 3, but applied to multiple assemblages. Subsampling level is tested. Presence of fish deformities, erosions, lesions, tumors (DELT) and other anomalies are quantified and documented.

Sample processing refers to the protocols (i.e., SOPs) that are followed to subsample, sort, identify, and count the organisms collected from a water body. These protocols include the specific methods for identifying organisms (e.g., by employing established keys), for training of the personnel who count and identify the organisms, and for QA/QC. Consistent protocols for sample processing can minimize the potential that differences in sample processing cause differences in site assessments.

Protocols for subsampling, including how the subsample is selected and how many organisms are counted should be specified. For most assemblages, it is infeasible to identify all the organisms in the sample, and, therefore, a subsample of the collected organisms is identified and counted. In general, the more organisms that are identified, the more accurately and precisely one can characterize the structure of the biological assemblage (e.g., Barbour and Gerritsen 1996; Ostermiller and Hawkins 2004; Cao and Hawkins 2005; Cao et al. 2007). However, sample processing costs increase with subsampling effort, so the relative benefits of increased subsampling effort versus processing costs should be considered and documented.

The most appropriate protocols can depend on the assemblage that is collected. For example, macroinvertebrates are more effectively sorted and identified in the laboratory (Nichols and Norris 2006), whereas fish are typically identified and counted in the field prior to returning them to the water body. (Note that when field identifications are used, voucher specimens should be retained for QA in the laboratory.) Similarly, the presence of deformities, erosions, lesions, and tumors (DELT) usually can only be assessed with fish samples.

Scores for this technical element are based on the degree to which sample processing is standardized, and the degree to which QA/QC procedures are both documented and implemented. Example evaluation questions are:

- Are standardized methods for sample processing in place?
- Do methods include processing macroinvertebrate and algae samples in the laboratory, retaining voucher specimens for fish, and using a formal QA/QC program?

33

- Is the increased accuracy and precision of more intense subsampling effort for macroinvertebrates and algae relative to the costs of subsampling documented?

- For fish, does the program record DELT and other anomalies?

Programs that score highly on this technical element process macroinvertebrate and algae samples in the laboratory, count DELT anomalies on fish, retain voucher specimens, and use a formal QA/QC program. The process used to select subsampling effort for macroinvertebrates and algal assemblages is documented, and it is sufficient for accurate and precise characterizations of assemblage structure.

Frequently Asked Question

Question: How does the level of macroinvertebrate subsampling affect the results of biological assessment?

Answer: In general, precision of site-specific estimates of taxon richness might improve with both sampling and subsampling effort. However, there may be diminishing returns for increasing subsample effort, and various studies have suggested that subsampling more than 500 macroinvertebrate organisms yields little or no additional precision or accuracy (e.g., Barbour and Gerritsen 1996; Ostermiller and Hawkins 2004; Cao and Hawkins 2005; Cao et al. 2007). The costs of increased sampling and subsampling effort at single sites needs to be considered in the overall program design with the information expected to be gained from more extensive sampling (increased number of sites and sample density). Depending on the questions to be answered, increased subsampling effort might increase precision and power for before-after and upstream-downstream investigations, while increased extent of sites might increase power for statewide status and trends investigations.

2.1.9 Data Management (Element 9)

(Lowest) 1.0	2.0	3.0	4.0 (Highest)
Sampling event data organized in a series of spreadsheets (e.g., by year, by data-type). QA/QC is cursory and mostly for transcription errors. Might be paper files only.	Databases for physical-chemical, and biological data, and geographic information exist (Access, dBase, Geographic Information System [GIS], etc.) but are not linked or integrated. Data-handling methods manuals are available. QA/QC for data entry, value ranges, and site locations. A documented data dictionary defines data fields in terms of field methods and data collection.	Relational databases that integrate all biological, physical, and chemical data (Oracle, SQL Server, Access, etc.). Validation checks that guard against inadvertently storing incorrect or incomplete sampling data. Fully documented and implemented QA/QC process. Structure provides for data export and analysis via query includes dedicated database management. Fully documented data dictionary. Access to all databases is available for routine analysis in support of condition assessment.	Same as Level 3 adding automated data review and validation tools. Numerous built-in data management and analysis tools to support routine and exploratory analyses. Ability to track history of changes made to the data. Ability to control who has privilege to change, update, or delete data. Data import and export tools. Integrated connection to GIS showing monitored sites in relation to other relevant spatial data layers. Fully documented metadata according to accepted database standards. Reports on commonly used endpoints are easily retrieved (e.g., menu driven).

The data management technical element evaluates the processes and systems that are used by a monitoring program to store and access collected data. A reliable, well-designed, and quality-assured database and management system is fundamental to a program's ability to effectively use monitoring information to assess environmental problems and allows historical data to be used to evaluate trends and provide historical context. Proper data management ensures that the appropriate data can be retrieved and analyzed when necessary and with ease of access, and that historical data are archived in a data repository to protect against data loss (e.g., Michener and Jones 2012).

Proper data management also requires documented metadata, that is, data about the data. Metadata documents are the who, what, why, where, when, and how of the data in the database, so it would include documentation of methods, units, design, objectives. The metadata ranges from methodological description of the study (or studies) to the data dictionary describing fields in the database. Metadata can be coded into Ecological Metadata Language, a metadata specification developed for ecology, based on work sponsored by the National Science Foundation (The Knowledge Network for Biocomplexity; http://knb.ecoinformatics.org/index.jsp).

Scoring of this technical element is based on the degree to which data management systems permit the program to retrieve data in formats that are useful for conducting analyses and supporting decision making. A low score in this element would be associated with simple spreadsheet storage of monitoring data. Higher scores would be associated with data stored in a relational database allowing integration with spatial data and providing stakeholders with Web access. Also, the methods used for archiving data and for making the data available to outside users are considered. Example evaluation questions are:

- Are data storage and analysis programs in place to access data, determine data quality, and manipulate the data to evaluate the relationship between measures of stressors or categories of stressors with biological assemblage response?

- Does data management include comprehensive and integrated storage of biological assessment, physical, chemical, WET, and watershed observations, such that these can be integrated with respect to space and time?

For a program to score high on this technical element, all monitoring data are stored in a relational database allowing integration with spatial data and providing users and stakeholders with Web access to access raw and summary data. Transparent and well-documented QA/QC procedures are in place for data storage and retrieval, including protocols for tracking changes in taxonomic nomenclature over time. All relevant data collected by the agency are in one integrated database system.

Frequently Asked Questions

Question: How do I know what type of data management system I need?

Answer: Data organization and management allows users to perform assessments and reorganize and summarize data according to analysis needs, including exploratory analyses, index development, and more advanced research. Use of spreadsheets is the minimum level of an electronic database management system, but spreadsheets are deficient in error checking and data integration, and they are limited in the amount of information that can be stored. A relational database addresses these shortcomings. A thorough QA/QC check on the database ensures a "clean" data set for use throughout an agency's program. A small relational database management system (RDBMS) such as Microsoft Access could serve as a logical step from spreadsheets to a more sophisticated relational database. These smaller systems can be used to develop a biological assessment database that includes most of the relational data integrity and validation features of a larger RDBMS. Most large RDBMS are installed on a server that provides options for making the database available through a network or Internet connection. Larger RDBMS are usually installed and administered by an agency's information technology (IT) department. IT departments can help program managers identify qualified professionals to assist with creating a custom database to meet the data management and analysis needs of biological assessment programs.

When developing a relational database, it is important to recognize that data access depends on creating and running queries, which must be properly programmed to extract appropriate data, and to make extracted data tables available to outside users as flat files.

Question: If I'm able to use electronic spreadsheets or even a small RDBMS such as Microsoft Access, why do I need a data dictionary (metadata)?

Answer: A well-documented data dictionary defines not only how the data in a particular field relate to field operations and data collection, but it specifies how those values are stored and validated. Creating a well-documented data dictionary requires the data manager to address questions ranging from fairly simple to more complex. For example, are the data numeric or text? Are they allowed to be null? The answers to these questions might show that multiple types of data are being stored in one field and should be separated. Answering these questions helps to bridge the gap between using spreadsheets and moving toward a more robust data management system.

2.1.10 Ecological Attributes (Element 10)

(Lowest) 1.0	2.0	3.0	4.0 (Highest)
Biological program relies solely on the evaluation of the presence or absence of targeted or key species. No rationale is provided for selection of indicators. Assessment endpoints and ecological attributes are not defined.	Biological program based on "off the shelf" indicators for one biological assemblage. Rationale for selection of indicators is partially documented. Generic assessment endpoints and ecological attributes are defined but not specifically evaluated for state or regional conditions.	Biological program based on well-developed ecological attributes for one biological assemblage. Rationale for attribute selection is thorough and well-documented. Explicit linkage is provided between management goal, assessment endpoints, and ecological attributes.	Same as Level 3, but biological program based on well-developed ecological attributes for two or more biological assemblages (e.g., faunal, flora) for more complete assessment of the members of an aquatic community.

The objective of the 1972 CWA is to "… to restore and maintain the chemical, physical and biological integrity of the Nation's waters." However, the CWA does not provide an explicit description of biological integrity nor specify ecological assessment endpoints and scientific methods to measure integrity. One description of biological integrity is "a balanced, integrated, and adaptive community of organisms having a composition and diversity comparable to that of natural habitats of the region" (Frey 1975; Karr and Dudley 1981). Primarily based on this definition or on later refinements (Karr and Chu 2000), states and tribes have used biological assessments to measure the condition of biological communities relative to biological integrity.

This technical element evaluates how well a biological assessment program has selected and operationally defined assessment endpoints that adequately represent biological integrity. Assessment endpoints are measurable characteristics, or attributes, representative of a management goal (USEPA 1998). The attributes provide the basis for development of quantitative measures (e.g., biological indices) to assess attainment of the management goal. Selection of attributes to measure biological integrity includes consideration of their ecological relevance, susceptibility to known or potential stressors, and relevance to the management goal (USEPA 1998). Ecologically relevant attributes might be identified at any level of organization (e.g., individual, population, community, ecosystem, landscape). Typically states and tribes have identified species diversity and abundance as ecologically relevant attributes for measuring biological integrity and have developed biological indices using measures of taxonomic diversity and completeness, composition, trophic state, and trophic composition.

Full consideration of all three selection criteria (e.g., ecological relevance, susceptibility to known or potential stressors, relevance to management goal) provides the best foundation for development of biological indices to measure biological integrity. Poorly defined attributes can lead to miscommunication and uncertainty in applying assessment results to making a judgment on attainment of the management goal. For example, susceptibility of an ecological attribute to stressors and/or levels of human disturbance in the environment is important in selecting attributes but should be considered in the context of how well an attribute can represent the management goal. Otherwise, an attribute could be selected that leads to a biotic index that provides a robust and precise measure of human disturbance but not an accurate measure of biological integrity.

Scientists from EPA, U.S. Geological Survey, state and tribal agencies, and academic institutions jointly developed a conceptual scientific model that describes the response of 10 ecological attributes to increasing anthropogenic stress (Davies and Jackson 2006, Table 2-3). This model, the Biological Condition Gradient (BCG), is based on a suite of ecological attributes used by different state and tribal biological assessment programs across the country. The BCG was developed to provide a common framework for interpretation of biological assessments regardless of methods or regional differences. The ecological attributes of the BCG might serve as a template, or starting point, for states and tribes to consider in their selection of attributes.

Scoring for this technical element is based on how a biological assessment program has selected and operationally defined ecological attributes to assess biological integrity and then used them as the basis for development of biological indices. Because the condition of a biological community can be more confidently assessed with more than one biotic assemblage, the number and type of assemblages are considered in the evaluation (e.g., Carlisle et al. 2008). Example evaluation questions are:

- Are ecological attributes defined that provide for development of biological indices to measure attainment of biological integrity? If so, what are the ecological attributes and what is the basis for their selection?
- What aquatic assemblages are assessed?
- How is the linkage between biological integrity, ecological attributes, and biological indices defined, tested, and documented?

Programs that receive the highest scores for this technical element have well-developed ecological attributes for two or more assemblages. The linkage between biological integrity, assessment endpoints, ecological attributes and the resulting biological indices is explicit and documented.

Table 2-3. Biological and other ecological attributes used to characterize the BCG

Attribute	Description
I. Historically documented, sensitive, long-lived, or regionally endemic taxa	Taxa known to have been supported according to historical, museum, or archaeological records, or taxa with restricted distribution (occurring only in a locale as opposed to a region), often due to unique life history requirements (e.g., sturgeon, American eel, pupfish, unionid mussel species).
II. Highly sensitive (typically uncommon) taxa	Taxa that are highly sensitive to pollution or anthropogenic disturbance. Tend to occur in low numbers, and many taxa are specialists for habitats and food type. These are the first to disappear with disturbance or pollution (e.g., most stoneflies, brook trout [in the east], brook lamprey).
III. Intermediate sensitive and common taxa	Common taxa that are ubiquitous and abundant in relatively undisturbed conditions but are sensitive to anthropogenic disturbance/pollution. They have a broader range of tolerance than highly sensitive taxa (attribute II) and can be found at reduced density and richness in moderately disturbed sites (e.g., many mayflies, many darter fish species).
IV. Taxa of intermediate tolerance	Ubiquitous and common taxa that can be found under almost any conditions, from undisturbed to highly stressed sites. They are broadly tolerant but often decline under extreme conditions (e.g., filter-feeding caddisflies, many midges, many minnow species).
V. Highly tolerant taxa	Taxa that typically are uncommon and of low abundance in undisturbed conditions but that increase in abundance in disturbed sites. Opportunistic species able to exploit resources in disturbed sites (e.g., tubificid worms, black bullhead).
VI. Nonnative or intentionally introduced species	Any species not native to the ecosystem (e.g., Asiatic clam, zebra mussel, carp, European brown trout). Additionally, there are many fish that have expanded their range within North America because they have been introduced to areas where they were not native.
VII. Organism condition	Anomalies of the organisms; indicators of individual health (e.g., deformities, erosions, lesions, tumors [DELT]).
VIII. Ecosystem function	Processes performed by ecosystems, including primary and secondary production; respiration; nutrient cycling; decomposition; their proportion/dominance; and what components of the system carry the dominant functions. For example, shift of lakes and estuaries to phytoplankton production and microbial decomposition under disturbance and eutrophication.
IX. Spatial and temporal extent of detrimental effects	The spatial and temporal extent of cumulative adverse effects of stressors, (e.g., widespread tile drainage and stream channelization throughout an ecoregion resulting in extirpation of several species of native macroinvertebrates and fish).
X. Ecosystem connectance	Access or linkage (in space/time) to materials, locations, and conditions required for maintenance of interacting populations of aquatic life; the opposite of fragmentation (e.g., levees restrict connections between flowing water and floodplain nutrient sinks [disrupt function]; dams impede fish migration and spawning).

Source: Modified from Davies and Jackson 2006.

Frequently Asked Questions

Question: Are all 10 BCG attributes necessary to characterize biological integrity?

Answer: The selection of attributes might depend on the spatial scale and specific water body being assessed. Each attribute provides some information about the biological condition of a water body. Combined into a conceptual model comparable to the BCG, the attributes can offer a more complete picture about current water body conditions and also provide a basis for comparison with naturally expected water body conditions. All states and tribes that have applied a BCG for streams, rivers, and wetlands have used the first seven attributes that describe the composition and structure of biotic community on the basis of the tolerance of species to stressors and, where available, included information on the presence or absence of native and nonnative species, and, for fish and amphibians, used measures of overall condition (e.g., size, weight, abnormalities, tumors). Though not measured directly in state or tribal stream biological assessment programs, the last three BCG attributes of ecosystem function and connectedness and spatial and temporal extent of stressors can provide valuable information when evaluating the potential for a stream, river, or wetland to be protected or restored. For example, a manager can choose to target resources and restoration activities to a stream where there is limited spatial extent of stressors or there are adjacent intact wetlands and stream buffers or intact hydrology, rather than a stream with comparable biological condition but where adjacent wetlands have been recently eliminated, hydrology altered, and stressor input is predicted to increase.

However, for comprehensive water body-wide assessments of large systems like estuaries and coastal ecosystems, the full suite of attributes might be important for application at both a single habitat scale similar to streams and for a landscape level assessment that describes the distribution and connectedness of habitats within an ecosystem necessary for the survival and resiliency of the resident biota (e.g., fish, benthic invertebrates, migratory water birds, aquatic mammals).

Question: I have a calibrated index. Why do I need to consider the ecological attributes of the BCG?

Answer: The BCG serves as a conceptual model, or framework, for organizing and communicating information on biological community response to increasing levels of stress in aquatic ecosystems. The BCG was developed in partnership with scientists from state and tribal biological assessment programs from across the country (Davies and Jackson 2006). The BCG attributes and levels of condition represent shared, measurable patterns of biological response to increasing stress condition regardless of location and method. Many of the state and tribal scientists involved in BCG development had already derived biological indices based on methods and approaches developed in the 1980s through 1990s (e.g., index of biotic integrity (IBI) for fish [Karr et al.1986]). Therefore, there is both conceptually and quantitatively a close association between BCG attributes and the biological indices currently used by many states and tribes. The suite of BCG attributes can serve as a template for reviewing and improving an existing biological index or for developing a new index.

Question: What is a trait-based approach?

Answer: A trait-based approach predicts patterns of species attributes (i.e., reproductive, physiological, behavioral) and environmental conditions (Poff et al. 2006; Pollard and Yuan 2010). This approach has not been consistently applied or formally articulated until the last decade. It is based on sound theoretical concepts, such as the Habitat Templet Concept, which predicts that habitat and environmental conditions select organisms with particular life-history strategies and biological traits (Southwood 1977, 1988). Many studies have demonstrated that patterns in the traits of species can be related to environmental conditions (e.g., Townsend et al. 1997; Richards et al. 1997; Statzner et al. 2005; Van Kleef et al. 2006).

2.1.11 Discriminatory Capacity (Element 11)

(Lowest) 1.0	2.0	3.0	4.0 (Highest)
Coarse method (low signal) and detects only high and low values. Supports distinguishing only extreme change in biological condition at the upper and lower ends of a generalized stress gradient.	A biological index for one assemblage is established but is not calibrated for water body classes, regional or statewide applications. BPJ based on single dimension attributes. The index can distinguish two general levels of change in biological condition along a generalized stress gradient.	A biological index for one assemblage has been developed and calibrated for statewide or regional application and for all classes and strata of a given water body type. The index can distinguish 3 to 4 increments of biological change along a continuous stress gradient. Supports narrative evaluations (e.g., good, fair, poor) based on multimetric or multivariate analyses that are relevant to the selected ecological attributes (Technical Element 10).	Same as Level 3 but biological indices for two or more assemblages have been developed and calibrated. Additionally, the indices can distinguish finer increments of biological change along a continuous stress gradient. The number of increments that potentially can be distinguished is dependent on water body type and natural climatic and geographic factors.

This technical element addresses how a biological assessment program has developed one or more biological indices based on ecological attributes (Technical Element 10) and the degree of sensitivity of the indices in distinguishing incremental change along a continuous gradient of stress. Detailed descriptions of biological change along a gradient of stress can provide detailed descriptions of a state's designated aquatic life uses for specific water bodies and regions and lead to biological criteria development. Additionally, depending on the sensitivity, or discriminatory capacity, of the index, the information can be used to help identify high-quality waters and establish incremental restoration goals for degraded waters.

The ability of a biological index to measure change along a continuous gradient of stress includes consideration of the appropriate scale for application of the index (e.g., a specific water body, class of water body, region, statewide) and defining, and wherever possible, quantifying overall variability and sources of uncertainty.

The BCG discussed in the preceding section (Technical Element 10) is a conceptual model that describes measurable increments of biological change along a gradient of stress (Davies and Jackson 2006). Six general increments of change have been described for each of the BCG's ecological attributes. The gradient ranges from natural, undisturbed conditions to severely degraded conditions caused by anthropogenic stresses. These incremental changes can serve as a template for developing biological indices that represent aspects of biological integrity and show a predictable, measurable response to increasing levels of stress.

Scoring of this technical element is based on the demonstrated ability of the biological index to detect increments of change along a continuous gradient of stress. Examples of evaluation questions are:

- Is the index developed and calibrated at the appropriate scale for its intended application?

- Is the index developed and verified by independent data sets?

- What is the sensitivity of the index to detect shifts in biological assemblages along a full gradient of anthropogenic stress?

- How well defined, quantified, and documented is overall variability and its sources?

- What biotic assemblages are assessed?

Programs that score highly on this technical element have well-developed indices for one or more assemblages and have demonstrated the ability of their indices to distinguish incremental levels of biological condition change along a continuous stressor gradient for specific water body types and regions. Sources of uncertainty are well defined and quantified. For a program to score at the highest level, well-developed biological indices for two or more assemblages are used for a more complete assessment of biological integrity.

Frequently Asked Questions

Question: Can an agency's existing biological index be refined rather than replaced to improve discriminatory capacity?

Answer: As a biological index is further developed, it can be recalibrated and compared with performance of the previous iteration to compare past and present results. Recalibration of an index or model should be considered, for example, when sample collection or processing protocols change; classification is refined; level of taxonomic identification is made more precise; or, the data set is substantially expanded to include longer time-series, stressor conditions, or reference characteristics. These technical improvements can influence discriminatory capacity of an index or model.

Developing a quantitative translation between the original and refined index might require a special study where samples are collected simultaneously using the two protocols (for methodological changes). For example, in New England, alternative sampling and index methods were run side-by-side at the same sites (Snook et al. 2007). For minor methodological changes (e.g., taxonomic level, sampling or subsampling effort), analysis could be performed on samples that are virtually reformatted to provide two samples reflecting each protocol. For example, if Chironomidae (midges) were previously identified at the family level, but are currently identified at the genus level, the identifications in new samples could be reset at family level for calculation of the old index. Then comparisons of old and new indices could be performed on the reformatted and complete samples, yielding old and new index scores that could be compared through regression or other analyses. This would allow prediction of one index from the other and comparison of the assessment thresholds.

Question: Are the same increments of measurement expected for all aquatic water body ecotypes or in all regions of the United States?

Answer: The number of increments that can be distinguished is dependent not only on the water body ecotype and natural climatic and geographic factors that define the assemblage characteristics, but the effect of anthropogenic stressors. For example, the sensitivity of an index developed for a forested, high-gradient stream might support distinguishing five to six increments of change along a continuous stressor gradient while an intermittent, seasonal, or desert stream might support only three increments. Some of this is due to inherent natural characteristics of the assemblages and some might be due to current limitations of science and practice.

2.1.12 Stressor Association (Element 12)

(Lowest) 1.0	2.0	3.0	4.0 (Highest)
No ability to develop relationships between biological responses and anthropogenic stress.	Site-specific paired biological and stressor samples for studies of an individual water body or a segment of a water body (e.g., a stream reach). Stress-response relationships are developed based on assemblage attributes at coarse level taxonomy (e.g., family for benthic macroinvertebrates). Information might be used on a case-by-case basis to inform a first order causal analysis.	Low spatial resolution for paired biological and stressor samples in time and space across the state at basin or sub-basin scale (e.g., HUC 4–8). Stress-response relationships developed for one assemblage using regression analysis. Taxonomy at level sufficient to detect patterns of response to stress (e.g., species or genus for benthic macroinvertebrates or periphyton, species for fish). Relational database supports basic queries. Information is frequently used to inform causal analysis. Reevaluation of stress-response relationships on an as-needed basis.	High spatial resolution for paired biological (including DELT anomalies and other indicators of organism health) and stressor samples in time and space across the state at watershed or subwatershed scales (e.g., HUC 10–12). Other data (e.g., watershed characteristics, land use data and information, flow regime, habitat, climatic data) are linked to field data for source identification. Stress-response relationships are fully developed for two or more assemblages, stressors, and their sources using a suite of analytical approaches (e.g., multiple regression, multivariate techniques). Relational database supports complex queries. Information is routinely used to inform causal analysis and criteria development. Ongoing evaluation of stress-response relationships and monitoring for new stressors is supported.

Stressor association refers to the use of biological assessment data at appropriate levels of taxonomy to develop relationships between measures of biological response and anthropogenic stressors, including both stressor and their sources (Yuan and Norton 2003; Huff et al. 2006; Yuan 2010; Miller et al., 2012). This includes examination of biological assessment data for patterns of response to categorical stressors (Yoder and Rankin 1995b; Riva-Murray et al. 2002; Yoder and DeShon 2003). A capability for developing these relationships extends the use of biological assessments from assessing condition to informing identification of possible causes and sources of a biological impairment at multiple scales.[4]

The technical capability to associate biological response with stressors and their sources affecting aquatic systems requires a comprehensive database that should include biological, chemical, physical, and WET data and information; detailed watershed and land use

---

[4] For more information about stressor identification, see EPA's Causal Analysis/Diagnosis Decision Information System website at: http://www.epa.gov/caddis.

information; locations of discharges; discharge monitoring; Geographic Information System (GIS) capability to assemble watershed and discharge information and relate them to the correct sampling sites, etc. Paired biological and other relevant environmental data support developing quantitative stress-response relationships. A relational database that enables data export and analysis via query is required to support this function. Since chemical sampling is often more frequent (several times per year) than biological sampling, the database should be able to accommodate queries to relate the higher-frequency chemical sampling to lower-frequency biological sampling. It should also be able to reveal the spatial coincidence of biological and chemical/physical sampling locations to reveal the extent to which these are actually paired.

Stressor association, is directly dependent on a high level of technical development of other elements, particularly the elements for spatial sampling design, taxa and level of taxonomic resolution, database management, and discriminatory capacity. These elements are important building blocks for the data collection and analysis needed to more confidently identify stressors and their sources and to estimate stress-response relationships. For example, the ability to estimate these relationships relies on paired stressor and response sampling at appropriate spatial and temporal scales and a level of taxonomic resolution and index sensitivity sufficient to detect incremental biological changes along a stress gradient. Also, a relational database that supports complex queries enables efficient and full utilization of data. A high level of technical development for each of these elements and others provides the foundation for stressor association.

Scoring for this technical element is based on the degree to which biological assessments are used to estimate stress-response relationships and discern patterns of response to individual or categorical stressors. Example evaluation questions are:

- Are biological sample collection and stressor sample collection coordinated? What assemblages are sampled and to what level of taxonomy?

- Does the database support analysis of biological responses to individual stressors or categories of stressors? If so, at which spatial scale(s)?

- Is a systematic approach for identifying stressors at biologically degraded sites used? Is this information used on a routine basis to support identification of probable cause of the biological impacts and source of the stressors?

- Does the database support the continued analysis of biological responses, including WET, to individual stressors or categories of stressors especially as additional data are collected and as stressors change over time?

Programs receiving the highest score on this technical element collect data and conduct analyses that enable the estimation of relationships between biological responses for two or more assemblages and the dominant stressors in their regions. Data sets are examined to discern patterns of response to categorical stressors and for source identification. To elucidate stress-response relationships, the biotic and abiotic data and measurements must be both temporally and spatially linked in data sets. Within-site variability is characterized and appropriately incorporated into the analysis. New monitoring data and information on changes in land use and new stressors are systematically gathered and evaluated as a part of the routine monitoring and assessment program so that new stressors and their biological impacts are detected and stressor-response variables developed accordingly. Information is used to inform causal analysis and support criteria development. Timely information is also provided to other water quality programs to meet their information needs on stressor-response relationships and causal analysis.

Frequently Asked Questions

Question: What biological assessment information can be used as a basis for diagnosing problems?
Answer: Appropriately detailed biological assessment information is needed to discriminate between different categories of stressors and requires analyses of large data sets to reveal patterns of biological response across spatial and temporal gradients. To further examine for patterns of biological response to stress, equally detailed information on stressors, habitat, potential sources, and the natural background condition are also needed.

Question: How does one analyze stress-response?
Answer: There is a large and growing base of literature exploring different approaches to analyzing stress-response relationships from field data. Methods range from simple regressions to complex multivariate models and new methodologies (see Legendre and Legendre 1998 for an overview). The objective is to find community-level diagnostics, also called biological response signatures, which are characteristics of a biological community and are associated with specific stressors or categories of stressors and can be used diagnostically. In some cases, these indicators have been used by agencies to identify possible stressors from biological data (Yoder and DeShon 2003; Yoder and Rankin 1995b; Riva-Murray et al. 2002). A further refinement to this approach compares stressor-specific tolerance values associated with taxa collected at sampling sites with those from an expected assemblage predicted by a RIVPACS-type model (Huff et al. 2006; Hubler 2008). Additionally, new analytical approaches are being explored for identifying patterns of biological response to individual stressors, types or categories of stressors, and/or their sources (e.g., Shipley 2000; USEPA 2000; Oksanen and Minchen 2002; Cade and Noon 2003; Cormier et al. 2008; Baker and King 2009; King and Baker 2010; USEPA 2010a; Cormier et al. 2013).

Question: What are biomarkers, and can they be used for diagnosis?
Answer: Biomarkers are histopathological or biochemical signatures found in organisms that indicate some combination of stress, exposure to specific chemicals, or a disease. They are typically assayed from single individuals, where several individuals from a single site are sampled. They have been used most often in attempts to diagnose causes of observed impairments or mortality in fish. For example, Ripley et al. (2008) examined protein expression profiles of smallmouth bass in the Shenandoah River to identify candidate causes of biological impairment of the river and of several fish kills. They found that fish in the Shenandoah are immunologically stressed; however, there are multiple candidate causes of the stress (eutrophication, pesticides, agricultural animal runoff) (Ripley et al. 2008). Biomarkers of exposure to polycyclic aromatic hydrocarbons (PAHs) were examined in fish in contaminated rivers in Ohio, and they were key in identification of PAHs as one of several causes of biological impairment in the rivers (Lin et al. 2001; Yoder and DeShon 2003). This example illustrates how biological assessments in combination with other biological, chemical, or physical information support more robust causal analysis.

2.1.13 Professional Review (Element 13)

(Lowest) 1.0	2.0	3.0	4.0 (Highest)
Review is limited to editorial aspects. No technical review.	Internal technical review only.	Outside review of documentation and reports are conducted on an ad hoc basis.	Formal process for technical review to include multiple reference and documented system for reconciliation of comments and issues. Process results in methods and reporting improvements. Can include production of peer-reviewed journal publications by the agency.

The professional review technical element is the level to which agency data, methods, and procedures are reviewed, especially with regard to external stakeholder and scientific peer reviews. Subjecting documented methods and assessment reports to rigorous scientific peer review is ultimately the best way to ensure that an agency's data and scientific underpinnings are credible. Inherently, scientific peer reviews should be conducted in an objective and independent manner (outside the agency and with no vested interest in the outcome) by technical and other experts able to provide valid critique and suggestions, and where recommendations for improvement and refinement are taken in good faith. Validation of SOPs for all aspects of the assessment and monitoring program by outside experts is an initial step in establishing confidence in the resulting data. Programs that do not address and implement critical recommendations fail to benefit from an independent endorsement of their procedures and assessments.

The scoring for this technical element is based on the level of scientific peer review. Example evaluation questions are:

- Are documented methods and assessment reports subject to a rigorous scientific peer review process?
- What type of peer review is conducted, and how does the agency address review comments and document its response?

To score high in this technical element, a program will have a formal process for routine scientific peer review of data and documents. Programs with a high level of rigor ensure that reviews are done by outside, independent reviewers. The agency will also have an established, transparent process for documenting and tracking how it responds to comments from reviewers. Technical approaches might be included in peer review journal articles.

2.2 Determining the Overall Technical Program Level of Rigor

A technical element's scoring matrix or "checklist" has been developed to rate or score the key technical elements according to a four-tiered narrative description along a sliding scale that ranges from 1 to 4 (Appendix E). The checklist is used to evaluate each element and rate it independently as part of the overall program evaluation process. The scoring of the individual technical elements is based on the role of each element in supporting a biological assessment program's ability to:

- Assess biological condition of a water body in terms of biological integrity.

- Define biological change along a gradient of stress.

- Relate biological response to stressors and develop stress-response relationships.

EPA recognizes that the components of the various technical elements are inherently interrelated and the status or refinement of one element can influence others. However, focusing on individual elements first and then aggregating them into a cumulative rating provides an estimate for the overall level of rigor of a biological assessment program. The individual technical element scores can be used to prioritize specific areas for corrective actions and improvement, and these are detailed in Appendix E. The checklist should be completed for major water body types (e.g., flowing waters, lakes,

☛ The 13 technical elements are evaluated equally for the purpose of identifying strengths and areas for improvement. Clearly, several entail greater level of effort for development. Many are building blocks for others. For example, Technical Element 5, Reference Condition, evaluates the number of reference sites that are available based on reference site section factors (Technical Element 4); the degree to which the reference sites represent natural environmental gradients (Technical Element 3) and whether the number of sites is sufficient to support statistical evaluation of condition and derivation of numeric biological criteria. Likewise, Technical Element 12, Stressor Association, is influenced by whether there is sufficient spatial resolution (Technical Element 2) and natural classification (Technical Element 3) to characterize both natural and stress gradients as well as number of assemblages used to measure aquatic life use and detect stress-response relationships (Technical Element 6). Fundamental to this element is an adequate data management system (Technical Element 9) so that data is readily accessible and can be manipulated for complex analysis. The relationships between the technical elements and level of effort and sequence for each are part of the discussion in development of recommendations and action plan.

wetlands) with the assemblages used for each water body type noted. Different levels of biological assessment rigor might be evident among the different water body types and assemblages sampled, which is important for the water quality agency to determine and reconcile for management purposes.

It is important that the determination of the level of rigor be done with care to avoid an erroneous classification of the program. The evaluation of each technical element and the overall level of rigor of a biological assessment program should be done with the direct input of the state or tribal manager, supervisor(s), and technical staff. Documentation about the biological assessment program will be needed to complete various aspects of the checklist. The checklist should be completed for each water body ecotype as appropriate for the natural classification framework (e.g., lake, flowing waters, wetland, and per ecological region or other classification factors such as elevation) that the water quality agency routinely monitors. It is possible that different levels of rigor are being implemented for the different water body ecotypes within the jurisdiction of the state or tribe. The overall program score provides an indication of a biological assessment program's capability to derive biological criteria, describe biological change along a gradient of stress and develop response-stress relationships (Table 2-4).[5]

Table 2-4. Scoring associated with technical element levels of rigor

Level of Rigor	CE Score	% CE Score[6]
4	49–52	≥ 93.2
3	43–48	≥ 81.7–93.1
2	34–42	≥ 66.4–81.6
1	13–33	24.0–66.3

The central tendency of a biological assessment program's technical capability for each technical element is evaluated to arrive at a score. A score for one element might end up as a 3.5 if its central tendency is comparable to the technical capabilities of Level 3 but it has some technical characteristics of a Level 4 program and none of Level 2. It is important to emphasize that the evaluation process is intended to guide program development building on existing technical capabilities and addressing the gaps revealed in the review, rather than being viewed as a report card.

Summing the individual scores of the 13 technical elements provides a raw score for the biological assessment program with a range of 13–52. This score is then converted to a percent score by dividing the raw CE score by 52. The thresholds for determining the four levels of rigor

---

[5] Because the overall score is the result of the summation of individual scores for the 13 separate elements, the overall score does not establish minimum expectations regarding a state's ability to make decisions in context of different CWA regulatory programs. At all levels of technical development, biological assessment information can be used to support water quality decisions.

[6] The percent CE score is calculated based on 0.5 increments between CE raw scores.

are based on an allowable deviation from the maximum cumulative score of 52 across all 13 elements (Table 2-5). These thresholds correspond with improved program capabilities to detect shifts in biological assemblages along a gradient of stress, more comprehensively assess the biotic community, detect the suite of stressors impacting the biota, and quantify stressor-response relationships. For Level 4, there is a 3-point deviation or departure, a 9-point departure for Level 3, and an 18-point departure for Level 2. Deviations greater than 18 result in a Level 1 assignment.

Table 2-5. Allowable deviation of technical elements scores for each of the four levels of rigor

Level of Rigor	Departure from maximum cumulative score
4	-3
3	-9
2	-18
1	greater than -18

The levels of rigor are based on departures across the 13 technical elements as opposed to a strictly linear interpretation across the four narrative descriptions of each element (e.g., 3 x 13 = 39 as the maximum score for Level 3, 2 x 13 = 26 as the maximum score for Level 2). As such, the delineations of the four levels are based on the aggregate degree of departure across all 13 elements and in recognition that the overall level of rigor is an aggregate reflection of all 13 elements combined. It also recognizes the scoring across the four element narratives as an ordinal gradient as opposed to rigid and discrete categories. Based on the pilot evaluations, state and tribal biological assessment programs might exhibit characteristics of adjacent categories—hence the sliding scoring scale in 0.5 point increments.

The pilot testing done with states in 2002–2004 and follow-up evaluations conducted with selected states through 2010 show a congruence between the level of rigor and the formal adoption of numeric biological criteria and refined aquatic life uses in WQS (Table 2-6). Of the three states that have adopted numeric biological criteria and/or refined aquatic life uses in their WQS, two are Level 4 programs and one is 0.5 point from Level 4. Of the remaining five Level 3 states, three were considering developing numeric biological criteria and refined aquatic life uses, and each was expecting to continue technical development towards Level 4 as a result of ongoing technical and program developmental efforts. For states either achieving or developing a Level 4 program, coordinated biological, WET, chemical, and physical assessments and implementation of stressor identification as part of the water quality management program were either in place or being planned for.

Table 2-6. State Pilot Biological Assessment Reviews: Correspondence of the level of rigor to adoption or development of refined aquatic life uses and/or biological criteria in state WQS

CE Level (n)	Refined Aquatic Life Uses &Biological Criteria in WQS[7]	Refined Aquatic Life Uses & Biological Criteria in Development	Not Developing Refined Aquatic life Uses &/or Biological Criteria in WQS
4 (2)	2		
3 (5)	1	3	2
2 (14)	0	0	14
1 (0)	0	0	0

The guiding principles of the technical elements approach are intended to help state and tribal monitoring and assessment programs achieve levels of standardization, rigor, reliability, and reproducibility that are reasonably attainable under current technology and available funding (Yoder and Barbour 2009). While the assignment of a biological assessment program to one of the four levels of rigor has meaning and utility as a summary tool for assessing overall progress, how a state or tribe responds to the evaluation results is the critical action. For Level 4 programs, the focus is on program maintenance and how the program is incorporating new advances in the science and technology of biological assessment. In contrast, for Level 1, 2 and 3 programs, the focus is on the technical developments that are either already underway or that need to take place to meet the agency's needs for biological assessment data and information.

---

[7] includes biologically-based refined uses only.

CHAPTER 3: THE PROGRAM EVALUATION PROCESS

3.1 Introduction to the Evaluation Process

The biological program review is a systematic process to evaluate the technical capabilities of a state's biological assessment program and to identify next steps for overall program improvement. In this process, an expert reviewer conducts in-person interviews with the water quality agency and guides discussions with water quality agency managers and staff. Regional U.S. Environmental Protection Agency (EPA) managers and/or staff typically participate in the review and provide support to the process. The number of water quality agency personnel engaged in the review usually varies depending on the topic of discussion. The biological assessment and Water Quality Standards (WQS) program managers and technical staff are present throughout the review and constitute the core technical review team. Managers and staff from other programs within the agency, as well as other state agencies that conduct biological monitoring and assessments, might participate for the full workshop or engage for specific topics, overall summary discussions, and the concluding session (see Figure 3-1).

The expert reviewer acts as a facilitator to provide an objective perspective on a state's biological assessment program and to lead the review process, including the scoring of the individual technical elements and writing the results (e.g., the technical memorandum). Important considerations for selection of an expert reviewer include:

- Expertise in biological assessments and aquatic ecology.

- In-depth experience in conducting biological assessments and data analysis.

- Practical and applied knowledge of state and tribal biological assessment programs.

- Ability to facilitate the review and complete the technical memorandum objectively.

The review is composed of two parts (Figure 3-1). The first part of the review provides an overview of the biological assessment program and involves discussion of many aspects of the biological assessment program and how that information is used by different water quality programs. The second part of the review, the technical elements review, is the evaluation by the core review team of the technical rigor of the biological assessment program. The first part of the review focuses on program background to provide context for a state or tribal water quality management program to evaluate the type and quality of biological assessments appropriate to answering specific information needs. Using the review results as a road map, a state or tribe can develop a technical program to support its intended use of biological assessments. This is why the first part of the review process includes discussion of how a program functions and whether the biological assessment program is providing the type and level of information needed by the state or tribe. This discussion sets the stage for the technical evaluation—the determination of biological assessment program strengths and limitations in context of an agency's water quality management program information needs.

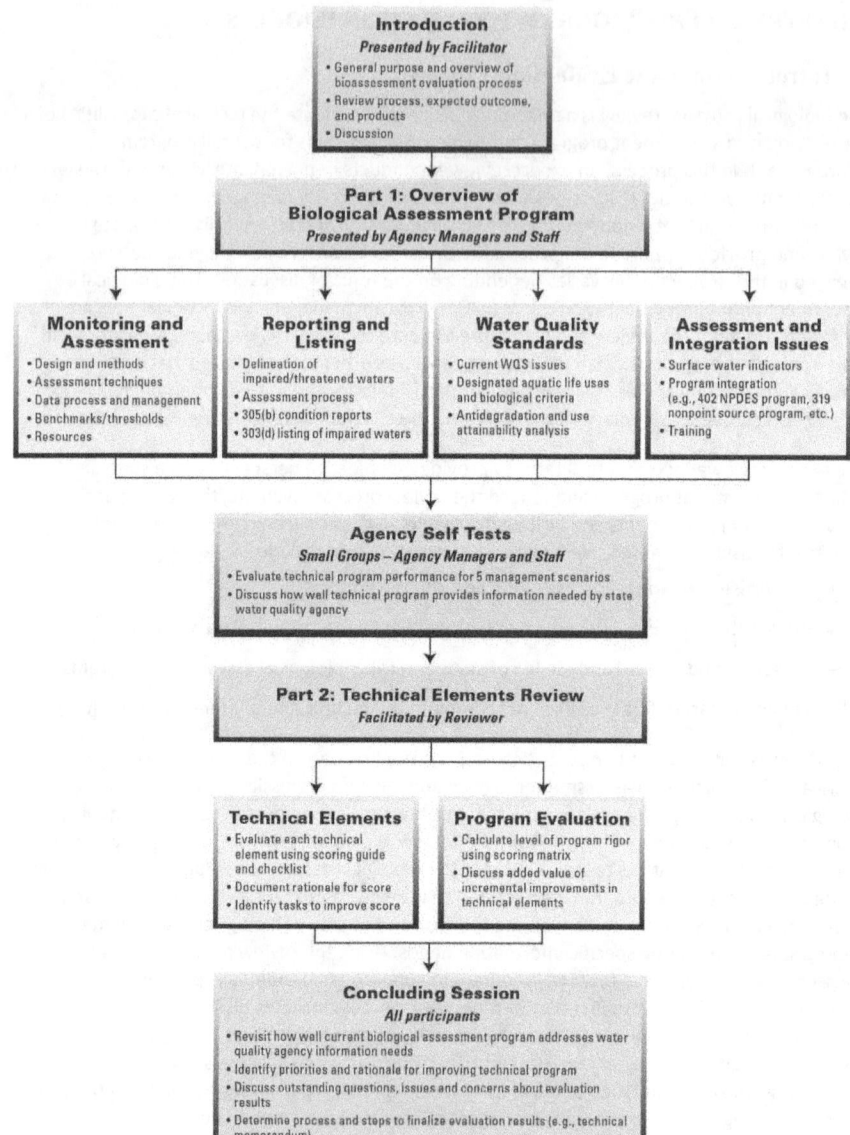

Figure 3-1. Flow chart of the 3-day biological assessment program evaluation process.

During the first part of the review—the overview—the reviewer leads the team in a discussion of the water quality agency's monitoring and assessment program, WQS and programs such as the Total Maximum Daily Load (TMDL), National Pollutant Discharge Elimination System (NPDES) permits, and nonpoint source programs. The discussion also serves as baseline fact finding for scoring each of the 13 technical elements of a biological assessment program and for identifying how the agency is currently using biological assessments and considering future applications (a complete listing of all annotated discussion topics is available in Appendix B: Interview Topics for Agency Review). This discussion provides managers and technical personnel with a better understanding of the program's history, why decisions were made, and how managers and staff interact across the monitoring and assessment program, WQS, listing, TMDL, NPDES, and nonpoint source programs. The discussion provides insight to the agency participants on the current technical strengths and deficits of the biological assessment program and the improvements needed to better support water quality management.

In the second part of the program review, the core review team evaluates 13 technical elements of a biological assessment program associated with biological assessment design, methods, and analysis. Through evaluation of the technical elements, the review team works together to assign a level of rigor (1–4) for the overall program based on the factors outlined in Chapter 2. On the basis of the discussion in the first part of the review, the review team develops a list of recommendations that the water quality agency can use to improve its program.

The final outcome of the program review is a technical memorandum written by the reviewer in collaboration with the full review team. In the memorandum, the reviewer describes important attributes of the overall program, summarizes the water quality agency's biological assessment program, justifies the assignment of the program's level or rigor, and recommends future actions. A step-by-step guide for conducting a biological assessment program evaluation is below.

3.2 Preparation for the Review

For a biological program review to be successful, preparation is necessary for the reviewer as well as the water quality agency personnel. Key tasks for the water quality agency include 1) identifying a comprehensive list of program managers and staff to attend the review; 2) communicating the importance and purpose of each person's participation; and 3) providing materials that the expert reviewer uses to become knowledgeable about the state program.

3.2.1 Identifying Participants

It is essential that water quality agency personnel from different program areas are engaged in the discussions so that data quality and information requirements are accurately represented and properly implemented, especially with regard to EPA published methodologies. Participation from different water quality programs, for example, is also important in the review to build a shared understanding and broad perspective on the existing use of biological assessment information and to begin to identify the technical program gaps and areas for improved use. One person from the water quality agency is designated as the lead for the effort. This state contact is responsible for bringing together the appropriate state personnel and ensuring that necessary documentation is compiled for the review.

Participants should include both agency managers and staff involved in the following programs:

- WQS
- Monitoring and assessment
- Reporting and listing
    - Section 305(b)/303(d) integrated report and listings
- TMDL development and implementation
- Planning
- Nonpoint source assessment and management
- Dredge and fill (section 404/401)
- NPDES program
- Other relevant programs

The reviewer will designate a member of the water quality agency review team to serve as a note taker. The note taker should be available for the entire evaluation and is responsible for ensuring that all discussion is captured. These notes will aid the reviewer with developing the technical memorandum.

3.2.2 Materials Provided as Basis for Program Review

This guidance document itself should be distributed to the water quality agency personnel prior to beginning the program review to provide participants with an understanding of the technical elements and the checklist process. The document also introduces the water quality agency to the next steps in the biological criteria implementation process, including the option for the water quality agency to develop a timeline for achieving a biological assessment program of Level 4 rigor by setting specific milestones for program development.

The appendices include the materials to be used during the evaluation and as supplemental information. By reviewing this chapter and appendices prior to the on-site visit, personnel can familiarize themselves with their content. Some of these documents serve simply as templates and are modified by the reviewer prior to the review.

- **Agenda (Appendix A)**—outlines the basic structure of a biological assessment program evaluation. It is conceptual in design, open to input from both the water quality agency and reviewer, and serves as a starting point for coordinators to plan the evaluation. A review-specific agenda is developed prior to the review itself.

- **Water Quality Agency Interview Topics (Appendix B)**—provides an overview of the major topics addressed during the biological assessment evaluation. The water quality agency is also encouraged to identify topic areas of interest and is free to steer the discussion accordingly. The reviewer and note taker each utilize this format for recording answers and discussion content.

- **Water Quality Agency Self-Assessments (Appendix C)**—designed to facilitate internal consideration about how the water quality agency's present biological assessment program can respond to specific water quality program information needs.

- **Technical Memorandum Template (Appendix D)**—serves as an example of the scope and content of the technical memorandum, the principal product of the biological assessment program evaluation.

- **Technical Elements Checklist (Appendix E)**—worksheet for evaluating the degree of development for each technical element of an agency's biological assessment program and associated comments on the elements for the biological assessment program.

3.2.3 Preparation of Documents

Prior to the review, the water quality agency compiles documentation that describes the state's decision-making process, the legal and regulatory framework, and technical components of the overall water quality management program (electronic links or documents are preferred). Access to the following materials should be provided to the independent expert reviewer prior to the site visit:

- Monitoring strategy
- WQS documents
- Biological standard operating procedures (SOPs)
- Listing methodology/guidance
- Section 305(b) report/303(d) list
- Example biological assessment reports/watershed assessments
- Any other materials the agency might determine relevant to the review, such as SOPs for other types of data (e.g., stressors, Geographic Information Systems [GIS])

The reviewer uses these materials to prepare for the interview and in developing the technical memorandum. The water quality agency also prepares an overview of its biological program that includes a brief history and a description of both current and planned program developments. The detail and mode of this presentation is left to the discretion of the water quality agency.

3.3 Part 1: Overview of Current Water Quality Program

3.3.1 Introduction and Overviews

(1) Participants
At the beginning of the evaluation, the water quality agency lead introduces managers and technical staff and briefly describes the purpose and scope of the biological assessment program review process. Individual personnel also offer detail about their specific roles with respect to the water quality agency's biological assessment program. The introductions provide an opportunity for the reviewer to become more familiar with the participants.

(2) Role of Biological Assessment
The reviewer begins the evaluation by giving a presentation to briefly introduce the key concepts of biological assessment-based aquatic life uses and biological criteria in relation to a water quality agency's biological monitoring and assessment program. The presentation, *Aquatic Life Uses: A Conceptual and Practical Basis for Determining Water Quality Management Goals and Outcomes Using Biological Assessments,* covers the relationships of biological, chemical, and physical indicators and criteria in the assessment of a water body's ecological health and the importance of using a system with which the biological response to stress in a water body can be evaluated. Topics included are:

- The linkage of biological assessments to other monitoring and assessment programs, with a focus on the WQS program.

- Information on how a biological assessment-based approach to water quality management support meeting the goals set forth by the water quality agency and Clean Water Act (CWA).

- Case examples of biological assessment programs that either currently achieve, or are building towards, high quality technical programs.

(3) Agency Objectives for Biological Assessment
The next step of the process is the water quality agency presenting an overview of its biological assessment program. This overview helps inform the assessment of the technical elements that follows by defining current technical components, use of the biological assessment information, and how the information produced aligns with managers' expectations and information needs. The water quality agency monitoring coordinator is asked to articulate how the water quality agency views the purpose, goals, and objectives of its monitoring program. This is helpful to have on record as it defines, in the water quality agency's own words, what the water quality agency wants to accomplish and how it intends to use information gathered from monitoring efforts. The water quality agency should include a brief history and any current developments or updates, but the remainder of the presentation's specifics is left up to the water quality agency. Personnel can develop an overview that is water quality agency- and program-specific by highlighting the key aspects that are self-identified as being of high importance.

3.3.2 Monitoring and Assessment

Monitoring and assessment includes the systematic collection of data from the environment and their subsequent analysis to allow assessments regarding attainment status, severity, and extent of impairments, stressor identification, and pollutant source identification. Monitoring and assessment is used to support the reporting requirements mandated by the CWA and other water quality agency efforts to characterize the status of water bodies and plan and implement restoration efforts. Discussion of current agency data quality objectives and measurement quality objectives (DQOs and MQOs, respectively) is a critical part of this discussion and documentation. In addition to specific agency objectives, it is useful to gather information on whether the agency aligns its monitoring program with, or directly feeds into, local and federal monitoring and assessments. When the agency personnel later conduct a self-assessment, the DQOs, MQOs, and other information will factor into this assessment and might be reviewed and revised as a consequence.

The following information is discussed during the evaluation:

- Spatial sampling design—The water quality program personnel describe the sampling design(s) employed by the water quality agency (e.g., how the water quality agency determines sampling locations, such as using a rotating basin approach, a probability-based approach, or via fixed stations). In addition, the water quality agency identifies the various water body types for which a monitoring and assessment program exists, as the design might vary among resource types.

- Index periods—The water quality agency clarifies whether a seasonal index period exists by indicator and/or assemblage and whether considerations are given for index periods during attenuated flows.

- Chemical/physical/whole effluent toxicity (WET) assessment—To clarify the design and logistics of the water quality agency's sampling regime (e.g., chemical, physical, WET), the agency personnel provide the reviewer with specifics regarding survey design, parameters and indicators, sampling frequency, sampled media (i.e., water, sediment, fish tissue), and the type of samples collected (e.g., grabs, composites). In addition, the group identifies goals of the sampling, such as characterizing ambient conditions, long-term trend assessments, and the determination of reference conditions. Finally, agency personnel provide the reviewer with information regarding laboratory support, specifically quality assurance/quality control (QA/QC) procedures and analytical costs.

- Reference condition—Agency personnel provide information on whether reference sites have been established, and if so, how many and for what period. The water quality agency provides additional detail about reference conditions, such as how reference is determined (e.g., reference site selection), and explanation of the spatial organization of reference sites and the degree to which these sites are stratified by landscape or other classification schemes and method for determining nonattainment of reference condition (i.e., membership or non-membership in a set of reference sites).

- Data processing and management—A relational database is essential to a highly rigorous biological assessment program. The water quality agency provides information on several technical elements related to data: (1) how biological, chemical, and physical data are stored and whether analysis can be conducted across multiple sampling types and data sets; (2) data management QA/QC procedures (including any documentation); and (3) the accessibility of these data to both agency personnel and outside parties.

- Basin assessments—The water quality agency responds to questions about the scale of basin assessments (e.g., using hydrologic unit code [HUC] units as a basis for expressing spatial scale), how basins are selected, the number of sites in a typical assessment unit (e.g., site density), and the number of basin assessments the water quality agency conducts each year. In addition, any stratifying factors are discussed, such as watershed area or stream order, flow, and the total number of sampling sites. Analysis of the data acquisition process culminates with a discussion of the study planning process to determine the level of integration, if any, of the various monitoring disciplines and interactions with water quality management programs. Finally, to garner an understanding of the assessment process, the sequence of data analysis and reporting will be determined and any logistical concerns identified.

- Monitoring strategy—The water quality agency provides the latest version of its monitoring strategy for review and responds to questions about the frequency of updates. Through discussion the reviewer will establish whether DQOs are clearly defined and evaluate the usefulness of the strategy to guide implementation of the monitoring program and to ensure use of the information to support water quality program information needs.

- Resources—The water quality agency provides specifics regarding the allocation of full time employees (FTEs), particularly how they are allocated to monitoring and assessment for each of the major scientific disciplines and the proportion of monitoring and assessment FTEs compared to those devoted to other water quality management programs. The water quality agency should provide an organizational table for the CWA components of the various programs at the staff level, and it should include any contracted resources. Finally, the water quality agency should identify current funding sources, any existing resource limitations, and what additional resources, if any, are needed.

3.3.3 Reporting and Listing (CWA sections 305[b] and 303[d]) and TMDLs

This part of the evaluation deals with the process of producing integrated CWA section 305(b) and 303(d) reports, which identify waters with impaired or threatened uses, and TMDLs. These reports are often used to delineate program priorities and allocate resources, and the information in these reports will help the reviewer make determinations about how its biological assessment program is used.

- Identification of waters with impaired or threatened uses—The water quality agency provides information on the procedures, protocols, and assessment methods for identifying waters with impaired or threatened uses. The water quality agency provides details on what data (biological, physical, and/or chemical) and methodology are used to determine aquatic life use impairments, and whether such impairments are based on assessment of aquatic life assemblages. Discussion can include the degree to which impairments are characterized for level of severity, extent, and cause. Finally, the water quality agency provides details on the extent to which the state's waters have been assessed and what percentage of the total waters this figure comprises.

- Data acquisition and management process—The water quality agency explains the process for making assessments of condition and status, including how the data and information is documented and quality controlled and protected against unauthorized changes. The water quality agency also describes requirements regarding any data acquired by outside organizations (e.g., volunteer groups, water collaboratives), such as admission requirements and accuracy determinations. Finally, the reviewer evaluates the water quality agency's legislation (if any) pertaining to data management.

- CWA section 303(d) list topics—The water quality agency should describe the extent to which biological assessment information has been used to identify waters with impaired or threatened uses, under which 305(b)/303(d) integrated reporting categories such waters are assigned, and how the information is used in the planning process for establishing TMDL development schedules as part of the 303(d) list submittal. The water quality agency should also describe and discuss any issues concerning the integration of biological information into one assessment methodology for both CWA section 305(b) and 303(d) reporting.

- CWA section 303(d) list and TMDL development and implementation topics—The water quality agency should describe the extent to which data from biological assessments and stressor identification evaluations are used in the development of TMDLs and the evaluation of their implementation. Finally, the reviewer will want to discuss any specific CWA section 303(d) or TMDL resource considerations.

3.3.4 Water Quality Standards

The WQS section of the review focuses on the development and integration of designated aquatic life uses and biological criteria in the state's WQS program. WQS are the basis for judging the effectiveness of water quality management programs. The water quality agency should provide all participants with a copy of the state's WQS during the evaluation, and the reviewer asks participants to refer to specific parts of the document as they become relevant during the discussion.

- General issues—The water quality agency describes the basis of the agency's WQS, such as how chemical water quality criteria are derived and whether site-specific criteria have ever been developed. The water quality agency describes its antidegradation policy and implementation procedures. The discussion should also include how the monitoring and assessment program is integrated with the WQS program.

- Designated uses—The water quality agency should provide a description of its aquatic life use designations and explain the process for assigning uses to water bodies. The reviewer will want the agency to describe any other special considerations, such as tributary rules and application of default uses. In addition, any triggers for re-designations should be described. The water quality agency should describe what it recognizes as waters meeting the CWA section 101(a)(2) goals.

- Use attainability analysis (UAA)—The water quality agency should explain its protocol for conducting a UAA and describe what data or information might initiate the process. Discussion of current technical issues or obstacles encountered when conducting UAAs can be included to help determine need for additional biological assessment information or other types of environmental data.

- Biological criteria—The water quality agency provides the reviewer with information to determine whether biological criteria have been developed and whether such criteria are narrative, numeric, or both. Secondly, participants describe habitat assessments and associated criteria, if applicable. The agency provides information to help the reviewer understand the linkage between biological criteria and aquatic life designated uses and how this information has been used to support water quality management programs.

3.3.5 Integration of Monitoring, Reporting, Standards, and Management

Integrating information gathered from monitoring and assessment efforts with other water quality management programs is integral to the overall program's effectiveness. The topics below are designed to assess the state's development, use, and integration of biological assessment information into water quality management programs.

- Indicators for surface waters—The water quality agency should describe its existing measures of the effectiveness of its water quality management programs. In addition, the agency should gauge the dependency of these indicators on monitoring data and identify the most important measures of water quality management program success.

- Program integration—The water quality agency explains how water quality management programs have relied on information gathered from ambient monitoring and assessment, focusing discussion on specific programs, including WQS, nonpoint source assessment and management, TMDLs, NPDES permitting, CWA section 404/401 dredge and fill permits, and any other important permitting and planning schemes. The agency should explain how data gathered via monitoring and assessments are viewed in context of their importance to application to other water quality management programs.

- Training—The water quality agency provides information on training of agency program personnel, including the depth of training and its frequency. In addition, the water quality agency clarifies whether such training is extended to outside entities affected by management programs.

3.3.6 Self-Assessments

During the on-site review, the water quality agency completes two self-assessments. In the self-assessments, the reviewer guides the water quality agency through discussion questions (see Appendix C) to discuss how its existing program would respond to given situations and to consider what additional technical capability would optimize its program capability and efficiency. Cross program discussion will foster a more complete understanding within the agency of whether the current biological assessment program is providing the needed data and information in the appropriate time frame to support multiple water quality programs and potentially identify areas where technical changes would enhance use of the data and better support agency water quality program goals and objectives.

The water quality agency is asked to modify the discussion questions prior to the on-site evaluation to make them as relevant and applicable as possible, including substituting any terminology (e.g., specific types of aquatic resources). Agency personnel proceed through each of the discussion questions and summarize how the programs currently incorporate biological assessment information to support their programs and develop recommendations for improvements. Agency personnel are encouraged to include comments describing each answer and specifics on how the current state program would respond to the discussion question. Upon completion, the reviewer collects the information and recommends and uses them to help develop recommendations for technical development of the biological assessment program to be included in the technical memorandum.

3.4 Part 2: Technical Elements Evaluation

Following a brief presentation regarding the technical elements evaluation process, the reviewer leads a discussion about the 13 technical elements (described in chapter 2). During this discussion participants provide input on scoring (see chapter 2 and Appendix E). Once a score has been assigned for each of the 13 elements, the numbers are tabulated and converted to a percentage that yields the agency's level of rigor. The water quality agency also provides information about any in-progress improvements to the biological assessment program that will result in the elevation of the score for specific technical elements.

3.4.1 Technical Elements of State Biological Assessment Programs: A Process to Evaluate Program Rigor and Comparability

The review typically begins with an overview presentation of the evaluation process. The presentation can include ways states and tribes can determine their current level of rigor and how to use this information to achieve specific milestones to improve the overall level of program rigor. The overview can also include examples of previous assessments, specifically those from the EPA regional pilots that were conducted annually during 2002–2008 (Yoder and Barbour 2009; this document). The presentation might also include general recommendations that were made to the pilot states and tribes, which prescribe implementing high-level biological assessment programs as a continual, iterative process involving the creation of regional working groups consisting of water quality agency staff and regional EPA personnel.

3.4.2 Technical Elements Checklist

As described in Chapter 2, the 13 technical elements checklist (see Appendix E) is used to assign a level of rigor to a water quality agency's biological assessment program. Agency personnel and the reviewer will discuss the basis for the scores using the checklist for each of the 13 elements. The reviewer will assign a preliminary score for each of the 13 elements and take notes regarding the score's justification and any ongoing water quality agency efforts and/or program developments that would affect the score. A tour of field and/or laboratory facilities might also be conducted during this portion of the review. Once each of the 13 elements has been scored, the results are tabulated and a score is assigned. These results are discussed by the review team and steps to address program gaps are identified. The score determines the level of rigor of an agency's biological assessment program. The water quality agency and reviewer will discuss the results of the technical elements exercise during the on-site visit and through follow-up conversations after the technical memorandum has been received and reviewed by the water quality agency.

3.5 Preparation of Technical Memorandum

The final output of the biological assessment program evaluation is the technical memorandum. Using the detailed information and documents provided by the water quality agency, the reviewer prepares a technical memorandum that summarizes the agency's biological assessment program, assigns the program a level of rigor, and justifies this assignment by providing the scoring's rationale. The technical memorandum includes recommendations on how the water quality agency can improve its biological assessment program and the development and use of numeric biological criteria, and on what steps it can take to achieve a higher level of rigor. These recommendations typically include enhancements relative to design, methodology, and execution of credible data.

Following completion of the technical memorandum, the reviewer submits it to the water quality agency and EPA regional staff for review and comment. Once the comments are received, they are incorporated into a final version. A template for the technical memorandum is available in Appendix D.

3.6 Action Plan Development

The ultimate goal of the biological program review is to produce the data and information needed by water quality agencies to strategically plan and allocate resources to develop and support a high-quality biological assessment program. In addition to evaluating the technical elements of a biological assessment program, identification of water quality program information needs (e.g., CWA section 303[d] listing, TMDLS, NPDES, nonpoint sources) and the flow of data from the monitoring program to the different water quality programs is an essential part of the evaluation. The program review produces technical recommendations for development of a high-quality biological assessment program and for effective use of the data and information that the technical program will generate.

In 2006 EPA Region 5 convened a region and state workshop on development of biological assessment and criteria programs. A central theme at the workshop was the importance of parallel efforts to:

- Establish early dialogue between management and technical staff to determine how high quality biological assessment information will be incorporated into the water management program. This dialogue is critical to ensure that the monitoring program plans for the design and production of data and information that will support water program information needs.

- Plan for the appropriate use of biological assessment information as the monitoring and assessment program's level of technical rigor increases. At all levels of technical development, biological assessment information can be used to support water quality decisions. The degree of confidence with which this can be done varies depending on the questions being addressed. The information produced by a program with a low level of rigor might be used to support screening for high-quality or severely degraded conditions (e.g., looking for "hot spots" that need immediate attention) and to identify water quality limited waters. Additionally, the biological assessment methods characteristic of a low level program might be used to support special studies as long as the degree of confidence (e.g., within site variability) is characterized and documented. As the level of program rigor is increased, more comprehensive and detailed condition assessments can be produced to further support CWA section 305(b) reporting and 303(d) listing decisions and report environmental outcomes from water quality management actions. As the state further develops and refines its biological assessment measures in conjunction with chemical, physical, WET, and landscape assessments, the monitoring and assessment program is increasingly able to provide information that contributes to stressor identification and development of attainable restoration targets.

Based on the discussions with the 23 program reviews done to date, the technical program needs to be developed within context of management needs and agency policy so that the information ultimately produced is used to support water quality management. For example, a biological assessment program with a high level of rigor might have the technical capability to develop biological measures sensitive to early changes in biological assemblages. The agency might consider incorporating these measures into its numeric biological criteria and refining its aquatic life uses to support protection of excellent and good conditions and implement preventive actions. In the pilot states where the dialogue between the monitoring program and the parts of the water program that use the data did not occur regularly, biological assessment information to support water quality management had not been fully realized.

3.7 Summary

The integration of rigorous biological assessments with other environmental data and assessments (e.g., chemical, WET, physical, landscape) is important for developing a comprehensive, data-driven but cost effective approach to support water quality management (USEPA 2011c). Despite advancements and successes in water quality management since the CWA was enacted, pollutants (e.g., pathogens, metals, nitrogen, and phosphorus pollution) continue to be major causes of water quality degradation. Additionally, the impact of other significant stressors, including habitat loss and fragmentation, hydrologic alteration, invasive species, and climate change, can be better understood using analytical tools and information that can operate at the ecosystem scale, such as biological assessments.

The biological assessment program review can be a first step toward identifying the specific actions a water quality agency can take to attain a rigorous biological assessment program. Additionally, an agency's overall ability to make management decisions is enhanced by using biological assessment to more precisely define designated aquatic life uses, develop numeric biological criteria, and associate biological response to chemical, physical, and landscape data (USEPA 2011c). The results of the review are intended to inform incremental technical development, future use refinements, and biological criteria derivation in context of sound scientific information and well-integrated monitoring and assessment information. For example, Minnesota's biological assessment program underwent a review in 2005 and then developed a plan with milestones to implement the review recommendations. The review process helped Minnesota Pollution Control Agency produce a detailed plan for technical program development to support refining the state's designated aquatic life uses and development of numeric biological criteria for streams and rivers.[8] Likewise, the California biological assessment program underwent a technical elements review in 2009. At the time of the review, California was already implementing a plan to develop its biological assessment program, but participation in the review process helped California align its program to the national elements framework. This helped California reinforce the importance of several key program elements (e.g., reference conditions, data management) and helped secure sustained management support. In 2009 the state initiated a public process to develop biological objectives (numeric biological criteria) for perennial streams and rivers.[9] This effort has included the development of guidance for selecting and evaluating candidate causes of biological impairment in different regions of the state, using the EPA's causal assessment process as a starting point. The biological objectives will be used to establish numeric scoring tools for measuring stream ecological integrity and define numeric thresholds needed to protect the state's designated aquatic life uses.

Aquatic life can vary from water body to water body. One major challenge in defining and assessing designated aquatic life uses is separating the natural variability that is a function of water body type and the ecological region from the variability that results from exposure to

---

[8]http://www.pca.state.mn.us/index.php/water/water-permits-and-rules/water-rulemaking/tiered-aquatic-life-use-talu-framework.html
[9] http://www.waterboards.ca.gov/plans_policies/biological_objective.shtml

stressors. Rigorous biological assessment programs can provide the detailed information required to more precisely define the expected biotic community for a water body and derive numeric biological criteria. By accounting for natural variability in aquatic systems, rigorous biological assessments can help reduce a source of uncertainty and error in water quality management. Additionally, in nature there is a continuous gradient of biological response to increasing exposure to stressors. A rigorous biological assessment program can support other agency water quality programs with the technical capability to discriminate levels of biological response along a stressor gradient to help identify and protect high-quality waters and set attainable restoration goals for degraded waters.

By conducting rigorous biological assessments in conjunction with chemical, WET, physical, and landscape data and assessments, more detailed relationships between the aquatic resource, stressor agents, and management actions can be developed. This means that an agency's biological assessment program can provide data and information for more than general status assessments as required by CWA section 305(b) and that can be used to inform impact assessments, studies, and investigations to support an agency's section 303(d) list, TMDL, NPDES permitting, and nonpoint source programs. Each of these programs relies on monitoring and assessment and the WQS programs to provide an accurate delineation of impairments and their associated causes, as well as determine attainment of specific requirements (e.g., criteria) on which calculations of water quality based limits are based.

The biological assessment program review process provides information and technical recommendations to the agency to further develop its technical rigor and to enhance program application. It is the agency's decision on when and how to implement the review results and recommendations for program improvements. Involvement of EPA staff in the review process is recommended to align agency efforts and resources to support the desired program development and foster agency partnerships. For example, regional EPA staff was involved throughout the Minnesota review and were instrumental is aligning EPA support and assistance. In California, strong and sustained support from regional EPA staff helped consolidate the state's biological assessment infrastructure development and enabled the state to rapidly develop the technical basis for the state's biological criteria. If an agency is interested in conducting a biological assessment program review, it is recommended that agency personnel contact EPA's regional or headquarters biological criteria program for further information and to plan a review.

REFERENCES CITED

Angradi, T.R., D.W. Bolgrien, T.M. Jicha, M.S. Pearson, B.H. Hill, D.L. Taylor, E.W. Schweiger, L. Shepard, A.R. Batterman, and M.F. Mofett. 2009. A bioassessment for mid-continent great rivers: The Upper Mississippi, Missouri, and Ohio (USA). *Environmental Monitoring and Assessment* 152(1):425–442.

Bailey, R.C., M.G. Kennedy, M.Z. Dervish, and R.M. Taylor. 1998. Biological assessment of freshwater ecosystems using a reference condition approach: Comparing predicted and actual benthic invertebrate communities in Yukon streams. *Freshwater Biology* 39:765–774.

Bailey, R.C., R.H. Norris, and T.B. Reynoldson. 2004. *Bioassessment of Freshwater Ecosystems Using the Reference Condition Approach*. Kluwer Academic Publishers, Dordrecht, The Netherlands.

Baker, M.E., and R.S. King. 2009. A new method for detecting and interpreting biodiversity and ecological community thresholds. *Methods in Ecology and Evolution* 1(1):25–37. British Ecological Society. doi: 10.1111/j.2041-210X.2009.00007.x.

Barbour, M.T., J. Gerritsen, G.E. Griffith, R. Frydenborg, E. McCarron, J.S. White, and M.L. Bastian. 1996. A framework for biological criteria for Florida streams using benthic macroinvertebrates. *Journal of the North American Benthological Society* 15(2):185–211.

Barbour, M.T., and J. Gerritsen. 1996. Subsampling of benthic samples: A defense of the fixed-count method. *Journal of the North American Benthological Society* 15(3):386–391.

Bollmohr, S., and R. Schulz. 2009. Seasonal changes of macroinvertebrate communities in a western cape river, South Africa, receiving nonpoint-source insecticide pollution. *Environmental Toxicology and Chemistry* 28(4):809–817.

Brooks, A.J., B.C. Chessman, and T. Haeusler. 2011. Macroinvertebrate traits distinguish unregulated rivers subject to water abstraction. *Journal of the North American Benthological Society* 30(2):419–435.

Bryce, S.A., E.P. Larsen, R.M. Hughes, and P.R. Kaufmann. 1999. Assessing relative risks to aquatic ecosystems: A mid-Appalachian case study. *Journal of the American Water Resources Association* 35:23–36.

Buss, D.F., and A.S. Vitorino. 2010. Rapid Bioassessment Protocols using benthic macroinvertebrates in Brazil: Evaluation of taxonomic sufficiency. *Journal of the North American Benthological Society* 29(2):562–571.

Cade, B.S., and B.R. Noon. 2003. A gentle introduction to quantile regression for ecologists. *Frontiers in Ecology and the Environment* 1(8):412–420.

Cao, Y., and C.P. Hawkins. 2005. Simulating biological impairment to evaluate the accuracy of ecological indicators. *Journal of Applied Ecology* 42:954–965.

Cao, Y., and C.P. Hawkins. 2011. The comparability of bioassessments: A review of conceptual and methodological issues. *Journal of North American Benthological Society* 30(3):680–701.

Cao, Y., C.P. Hawkins, D. Larsen, and J. Van Sickle. 2007. Effects of sample standardization on mean species detectabilities and estimates of relative differences in species richness among assemblages. *The American Naturalist* 170(3):381–395.

Carlisle, D.M., C.P. Hawkins, M.R. Meador, M. Potapova, and J. Falcone. 2008. Biological assessments of Appalachian streams based on predictive models for fish, macroinvertebrate, and diatom assemblages. *Journal of the North American Benthological Society* 27(1):16–37.

Chessman, B.C., and M.J. Royal. 2004. Bioassessment without reference sites: Use of environmental filters to predict natural assemblages of river macroinvertebrates. *Journal of the North American Benthological Society* 23(3):599–615.

Clarke, R.T., M.T. Furse, J.F. Wright, and D. Moss. 1996. Derivation of a biological quality index for river sites: Comparison of the observed with the expected fauna. *Journal of Applied Statistics* 23(2–3):311–332.

Clements, W.H., D.M. Carlisle, J.M. Lazorchak, and P.C. Johnson. 2000. Heavy metals structure benthic communities in Colorado mountain streams. *Ecological Applications* 10(2):626–638.

Cormier, S.M., J.F. Paul, R.L. Spehar, P. Shaw-Allen, W.J. Berry, and G.W. Suter, II. 2008. Using field data and weight of evidence to develop water quality criteria. *Integrated Environmental Assessment and Management* 4(4):490–504.

Cormier, S.M., G.W. Suter, II, L. Zheng, and G.J. Pond. 2013. Assessing causation of the extirpation of stream macroinvertebrates by a mixture of ions. *Environmental Toxicology and Chemistry* 32(2):277–287.

Danielson, T.J., C.S. Loftin, L. Tsomides, J.L. DiFranco, and B. Connors. 2011. Algal bioassessment metrics for wadeable streams and rivers of Maine, USA. *Journal of the North American Benthological Society* 30(4):1033–1048

Danielson, T.J., C.S. Loftin, L. Tsomides, J.L. DiFranco, B. Connors, D.L. Courtemanch, F. Drummond, and S.P. Davies. 2012. An algal model for predicting attainment of tiered biological criteria of Maine's streams and rivers. *Freshwater Science* 31(2):318–340.

Davies, S.P., and S.K. Jackson. 2006. The biological condition gradient: A descriptive model for interpreting change in aquatic ecosystems. *Ecological Applications* 16(4):1251–1266.

Diamond, J., J.B. Stribling, L. Huff, and J. Gilliam. 2012. An approach for determining bioassessment performance and comparability. *Environmental Monitoring and Assessment* 184:2247–2260.

Dodds, W.K., and R.M. Oakes. 2004. A technique for establishing reference nutrient concentrations across watersheds affected by humans. *Limnology and Oceanography: Methods* 2:333–341.

Dyer, S.D., and X. Wang. 2002. A comparison of stream biological responses to discharge from wastewater treatment plants in high and low population density areas. *Environmental Toxicology and Chemistry* 21(5):1065–1075.

Eagleson, K.W., D.L. Lenat, L.W. Ausley, and F.B. Winborne. 1990. Comparison of measured instream biological responses with responses predicted using the *Ceriodaphnia dubia* chronic toxicity test. *Environmental Toxicity and Chemistry* 9:1091–1028

Fausch, D.O., J.R. Karr, and P.R. Yant. 1984. Regional application of an index of biotic integrity based on stream fish communities. *Transactions of the American Fisheries Society* 113:39–55.

Feio, M.J., T.B. Reynoldson, and M.A.S. Graça. 2006. The influence of taxonomic level on the performance of a predictive model for water quality assessment. *Canadian Journal of Fisheries and Aquatic Sciences* 63(2):367–376.

Fennessy, M., A. Jacobs, and M. Kentula. 2007. An evaluation of rapid methods for assessing the ecological condition of wetlands. *Wetlands* 27(3):543–560.

Flinders, C.A., R.L. Ragsdale, and T.J. Hall. 2009. Patterns of fish community structure in a long-term watershed-scale study to address the aquatic ecosystem effects of pulp and paper mill discharges in four US receiving streams. *Integrated Environmental Assessment and Management* 5(2):219-233.

Frey, D. 1975. Biological integrity of water: An historical approach. In *The Integrity of Water*, ed. R.K. Ballentine, and L.J. Guarraia, pp. 127–140. Proceedings of a Symposium, March 10–12, 1975. U.S. Environmental Protection Agency, Washington, DC.

Furse, M.T., D. Moss, J.F. Wright, and P.D. Armitage. 1984. The influence of seasonal and taxonomic factors on the ordination and classification of running-water sites in Great Britain and on the prediction of their macroinvertebrate communities. *Freshwater Biology* 14:257–280.

Gerritsen, J., J. Burton, and M.T. Barbour. 2000. *A Stream Condition Index for West Virginia Wadeable Streams*. Tetra Tech, Inc., Owings Mills, MD.

Gerth, W.J., and A.T. Herlihy. 2006. Effect of Sampling Different Habitat Types in Regional Macroinvertebrate Bioassessment Surveys. *Journal of the North American Benthological Society* 25(2):501–512.

Growns, I. 2009. Differences in bioregional classifications among four aquatic biotic groups: Implications for conservation reserve design and monitoring programs. *Journal of Environmental Management* 90(8):2652–2658. doi:10.1016/j.jenvman.2009.02.002.

Hawkins, C.P. 2006. Quantifying biological integrity by taxonomic completeness: Its utility in regional and global assessment. *Ecological Applications* 16(4):1277–1294.

Hawkins, C.P., and R.H. Norris. 2000a. Performance of different landscape classifications for aquatic bioassessments: Introduction to the series. *Journal of the North American Benthological Society* 19(3):367–369.

Hawkins, C.P., and R.H. Norris. 2000b. Effects of taxonomic resolution and use of subsets of the fauna on the performance of RIVPACS-type models. Pages 217-228 in J.F. Wright, D.W Sutcliffe, and M.T. Furse (Eds) *Assessing the Biological Water Quality of Freshwaters: RIVPACS and Similar Techniques*. Freshwater Biological Association, Ambleside, UK.

Hawkins, C.P., R.H. Norris, J. Gerritsen, R.M. Hughes, S.K. Jackson, R. H. Johnson, and R.J. Stevenson. 2000a. Evaluation of landscape classifications for biological assessments of freshwater ecosystems: Synthesis and recommendations. *Journal of North American Benthological Society* 19:541–556.

Hawkins, C.P., R.H. Norris, J.N. Hogue, and J.W. Feminella. 2000b. Development and evaluation of predictive models for measuring the biological integrity of streams. *Ecological Applications* 10(5):1456–1477. doi:10.1890/1051-0761(2000)010[1456:DAEOPM]2.0.CO;2.

Hawkins, C.P., J.R. Olson, and R.A. Hill. 2010. The reference condition: predicting benchmarks for ecological and water-quality assessments. *Journal of North American Benthological Society* 29(1):312–343.

Hawkins, C.P., and M.R. Vinson. 2011. *Weak correspondence between landscape classifications and stream invertebrate assemblages: Implications for bioassessment*. The Society for Freshwater Science. <http://www.jnabs.org/doi/abs/10.2307/1468111>. Accessed February 2013.

Herlihy, A.T., D.P. Larsen, S.G. Paulsen, N.S. Urquhart, and B.J. Rosenbaum. 2000. Designing a spatially balanced, randomized site selection process for regional stream surveys: The EMAP mid-Atlantic pilot study. *Environmental Monitoring and Assessment* 63(1):95–113.

Herlihy, A.T., S.G. Paulsen, J.V. Sickle, J.L. Stoddard, C.P. Hawkins, and L.L. Yuan. 2008. Striving for consistency in a national assessment: The challenges of applying a reference-condition approach at a continental scale. *Journal of the North American Benthological Society* 27: 860–877.

Hitt, N.P., and P.L. Angermeier. 2011. Fish community and bioassessment responses to stream network position. *Journal of the North American Benthological Society* 30(1):296–309.

Hubler, S. 2008. *PREDATOR: Development and Use of RIVPACS-type Macroinvertebrate Models to Assess the Biotic Condition of Wadeable Oregon Streams*. DEQ08-LAB-0048-TR. Oregon Department of Environmental Quality, Hillsboro, OR.

Huff, D.D., S. Hubler, Y. Pan, and D. Drake. 2006. *Detecting Shifts in Macroinvertebrate Community Requirements: Implicating Causes of Impairment in Streams*. DEQ06-LAB-0068-TR. Oregon Department of Environmental Quality, Hillsboro, OR.

Hughes, R.M., and D.V. Peck. 2008. Acquiring data for large aquatic resource surveys: The art of compromise among science, logistics, and reality. *Journal of the North American Benthological Society* 27(4):837–859.

Hughes, R.N., D.P. Larsen, and J.M. Omernik. 1986. Regional reference sites: A method for assessing stream potentials. *Environmental Management* 10(5):629–635.

Hynes, H.B.N. 1970. *The Ecology of Running Waters*. University of Toronto Press, Toronto, Canada.

Jackson, L.E., J.C. Kurtz, and W.S. Fisher. 2000. *Evaluation Guidelines for Ecological Indicators*. EPA/620/R-99/005. U.S. Environmental Protection Agency, Office of Research and Development, Research Triangle Park, NC.

Joy, M.K., and R.G. Death. 2002. Predictive modeling of freshwater fish as a biomonitoring tool in New Zealand. *Freshwater Biology* 47:2261–2275.

Kaller, M.D., and K.J. Hartman. 2004. Evidence of a threshold level of fine sediment accumulation for altering benthic macroinvertebrate communities. *Hydrobiologia* 518:95–104.

Kang, I.J., H. Yokota, Y. Oshima, Y. Tsuruda, T. Yamaguchi, M. Maeda, N. Imada, H. Tadokoro, and T. Honjo. 2002. Effect of 17-B estradiol on the reproduction of Japanese medaka (*Oryziaslatipes*). *Chemosphere* 47(1):71–80.

Karr, J.R., and E.W. Chu. 1999. Restoring Life. In *Running Waters: Better Biological Monitoring*. Island Press, Washington, DC.

Karr, J.R., and E.W. Chu. 2000. Sustaining living rivers. *Hydrobiologia* 422:1–14.

Karr, J.R., and D.R. Dudley. 1981. Ecological Perspectives on water quality goals. *Environmental Management* 5:55–68.

Karr, J.R., K.D. Fausch, P.L. Angermeier, P.R. Yant, and I.J. Schlosser. 1986. *Assessing Biological Integrity in Running Waters: A Method and its Rationale*. Special publication 5. Illinois Natural History Survey.

Kelly, M.G. 1998. Use of the trophic diatom index to monitor eutrophication in rivers. *Water Research* 32:236–242.

Kilgour, B.W., and L.W. Stanfield. 2006. Hindcasting reference conditions in streams. *American Fisheries Society Symposium* 48:623–639.

King, R.S., and M. E. Baker. 2010. Considerations for analyzing ecological community thresholds in response to anthropogenic environmental gradients. *Journal of the North American Benthological Society* 29(3):998–1008.

Kosnicki, E., and R.W. Sites. 2011. Seasonal predictability of benthic macroinvertebrate metrics and community structure with maturity-weighted abundances in a Missouri Ozark stream, USA. *Ecological Indicators* 11(2):704–714.

Kurtz, J.C., L.E. Jackson, and W.S. Fisher. 2001. Strategies for evaluating indicators based on guidelines from the Environmental Protection Agency's Office of Research and Development. *Ecological Indicators* 1:49–60.

Legendre, P., and L. Legendre. 1998. *Numerical Ecology*. Second English Edition. Elsevier, Amsterdam, NL.

Lenat, D.R., and V.H. Resh. 2001. Taxonomy and stream ecology—The benefits of genus- and species-level identifications. *Journal of the North American Benthological Society* 20:287–298.

Lin, E.L.C., T.W. Neiheisel, J. Flotemersch, B. Subramanian, D.E. Williams, M.R. Millward, and S.M. Cormier. 2001. Historical monitoring of biomarkers of PAH exposure of brown bullhead in the remediated Black River and the Cuyahoga River, Ohio. *Journal of Great Lakes Research* 27(2):191–198.

Linke, S., R.C. Bailey, and J. Schwindt. 1999. Temporal variability of stream bioassessments using benthic macroinvertebrates. *Freshwater Biology* 42(3):575–584.

Lovy, J., D.J. Speare, and G.M. Wright. 2007. Pathological effects caused by chronic treatment of rainbow trout with indomethacin. *Journal of Aquatic Animal Health* 19(2):94–98.

Marchant, R., A. Hirst, R.H. Norris, and L. Metzeling. 1999. Classification of macroinvertebrate communities across drainage basins in Victoria, Australia: Consequences of sampling on a broad spatial scale for predictive modeling. *Freshwater Biology* 41:253–268.

McCord, S.B., and P.R. Lambrecht. 2006. Seasonal succession in the aquatic insect community of an Ozark stream. *Journal of Freshwater Ecology* 21(2):323–329.

McCormick, P.V., and J. Cairns. 1994. Algae as indicators of environmental change. *Journal of Applied Phycology* 6(5–6):509–526.

McCormick, F.H., R.M. Hughes, P.R. Kaufmann, D.V. Peck, J.L. Stoddard, and A.T. Herlihy. 2001. Development of an index of biotic integrity for the Mid-Atlantic Highlands region. *Transactions of the American Fisheries Society* 130:857–877.

McElravy, E.P., G.A. Lamberti, and V.H. Resh. 1989. Year-to-year variation in the aquatic macroinvertebrate fauna of a Northern California USA stream. *Journal of the North American Benthological Society* 8:51–63.

Michener, W.K., and M.B. Jones. 2012. Ecoinformatics: Supporting ecology as a data-intensive science. *Trends in Ecology and Evolution* 27(2):86–93.

Miller, M.P., J.G. Kennen, J.A. Mabe, and S.V. Mize. 2012. Temporal trends in algae, benthic invertebrate, and fish assemblages in streams and rivers draining basins of varying land use in the south-central United States, 1993–2007. *Hydrobiologia* 684(1)15–33.

Moore, M.J.C, H.A. Langrehr, and T.R. Angradi. 2012. A submersed macrophyte index of condition for the Upper Mississippi River. *Ecological Indicators* 13(1): 196–205.

Moss, D., M.T. Furse, J.F. Wright, and P.D. Armitage. 1987. The prediction of the macro-invertebrate fauna of unpolluted running-water sites in Great Britain using environmental data. *Freshwater Biology* 17:41–52.

Moyle, P.B., and P.J. Randall. 1998. Evaluating the biotic integrity of watersheds in the Sierra Nevada, California. *Conservation Biology* 12:1318–1326.

Munné, A., and N. Prat. 2011. Effects of Mediterranean climate annual variability on stream biological quality assessment using macroinvertebrate communities. *Ecological Indicators* 11(2):651–662.

Nelson, J.S., E.J. Crossman, H. Espinosa-Pérez, L.T. Findley, C.R. Gilbert, R.N. Lea, and J.D. Williams. 2004. *Common and Scientific Names of Fishes from the United States Canada and Mexico*. 6th ed. Special Publication 29. American Fisheries Society, Bethesda, MD.

NRC (National Research Council). 2002. *Opportunities to Improve the U.S. Geological Survey National Water Quality Assessment Program*. National Research Council, Water Science and Technology Board, National Academies Press, Washington, DC.

Nichols, S.J., and R.H. Norris. 2006. River condition assessment may depend on the sub-sampling method: Field live-sort versus laboratory sub-sampling of invertebrates for bioassessment. *Hydrobiologia* 572:195–213.

NYSDEC (New York State Department of Environmental Conservation). 2009. *Standard Operating Procedure: Biological Monitoring of Surface Waters in New York State*. New York State Department of Environmental Conservation, Division of Water. <http://www.dec.ny.gov/docs/water_pdf/sbusop2009.pdf>. Accessed September 2012.

Oberdorff, T., D. Pont, B. Hugueny, and J.P. Porcher. 2002. Development and validation of a fish-based index for the assessment of 'river health' in France. *Freshwater Biology* 47:1720–1734.

Ohio EPA (Ohio Environmental Protection Agency). 2004. Biological and Water Quality Study of Big Darby Creek and Selected Tributaries 2001/2002. Logan, Champaign, Union, Madison, Franklin, and Pickaway Counties, Ohio. Technical Report EAS/2004-6-3. Ohio Environmental Protection Agency, Division of Surface Water, Columbus, OH. <http://www.epa.ohio.gov/dsw/document_index/psdindx.aspx>. Accessed January 2013.

Ohio EPA (Ohio Environmental Protection Agency). 2012. *Ohio Primary Headwater Habitat Streams*. Ohio Environmental Protection Agency. <http://www.epa.ohio.gov/dsw/wqs/headwaters/index.aspx>. Accessed January 2013.

Oksanen, J., and P.R. Minchin. 2002. Continuum theory revisited: What shape are species responses along ecological gradients? *Ecological Modeling* 157(2–3):119–129.

Olsen, A.R., and D.V. Peck. 2008. Survey design and extent estimates for the Wadeable Streams Assessment. *Journal of the North American Benthological Society* 27:822–836.

Olsen, A.R., J. Sedransk, D. Edwards, C.A. Gotway, W. Liggett, S. Rathbun, K.H. Reckhow, and L.J. Young. 1999. Statistical issues for monitoring ecological and natural resources in the United States. *Environmental Monitoring and Assessment* 54:1–54.

Olsen, A.R., B.D. Snyder, L.L. Stahl, and J.L. Pitt. 2009. Survey design for lakes and reservoirs in the United States to assess contaminants in fish tissue. *Environmental Monitoring and Assessment* 150:91–100.

Ostermiller, J.D., and C.P. Hawkins. 2004. Effects of sampling error on bioassessments of stream ecosystems: Application to RIVPACS-type models. *Journal of North American Benthological Society* 23(2):363–382.

Omernik, J.M. 1987. Ecoregions of the conterminous United States. *Annals of the Association of American Geographers* 77(1):118–125. doi:10.1111/j.1467-8306.1987.tb00149.x.

Pan, Y., R.J. Stevenson, B.H. Hill, A.T. Herlihy, and G.B. Collins. 1996. Using diatoms as indicators of ecological conditions in lotic systems: A regional assessment. *Journal of the North American Benthological Society* 15:481–495.

Parsons, M., and R. Norris. 1996. The effect of habitat-specific sampling on biological assessment of water quality using a predictive model. *Freshwater Biology* 36(2):419–434.

Patrick, R. 1949. A proposed biological measure of stream conditions based on a survey of the Conestoga Basin, Lancaster County, Pennsylvania. In *Proceedings of the Academy of Natural Sciences, Philadelphia* 101:277–341.

Petty, J.T., J.B. Fulton, M.P. Strager, G.T. Merovich, Jr., J.M. Stiles, and P.F. Ziemiewicz. 2010. Landscape indicators and thresholds of stream ecological impairment in an intensively mined Appalachian watershed. *Journal of the North American Benthological Society* 29(4):1292–1309.

Poff, N.L. 1997. Landscape filters and species traits: Towards mechanistic understanding and prediction in stream ecology. *Journal of the North American Benthological Society* 16(2):391. doi:10.2307/1468026.

Poff, N.L., J.D. Olden, N.K.M. Vieira, D.S. Finn, M.P. Simmons, and B.C. Kondratieff. 2006. Functional trait niches of North American lotic insects: Traits-based ecological applications in light of phylogenetic relationships. *Journal of the North American Benthological Society* 25(4):730–755.

Pollard, A.I., and L.L. Yuan. 2010. Assessing the consistency of response metrics of the invertebrate benthos: A comparison of trait- and identity-based measures. *Freshwater Biology* 55(7):1420–1429.

Ponader, K.C., D.F. Charles, T.J. Belton, and D.M. Winter. 2008. Total phosphorus inference models and indices for coastal plain streams based on diatom assemblages from artificial substrates. *Hydrobiologia* 610:139–152.

Pond, G.J., M.E. Passmore, F.A. Borsuk, L. Reynolds, and C.J. Rose. 2008. Downstream effects of mountaintop coal mining: Comparing biological conditions using family- and genus-level macroinvertebrate bioassessment tools. *Journal of the North American Benthological Society* 27(3):717–737.

Pond, G.J., J.E. Bailey, B.M. Lowman, and M.J. Whitman. 2012. Calibration and validation of a regionally and seasonally stratified macroinvertebrate index for West Virginia wadeable streams. *Environmental Monitoring and Assessment* (in press).

Potapova, M., and D.F. Charles. 2003. Distribution of benthic diatoms in U.S. rivers in relation to conductivity and ionic composition. *Freshwater Biology* 48:1311–1328.

Rabeni, C.F., and K.E. Doisy. 2011. Correspondence of stream benthic invertebrate assemblages to regional classification schemes in Missouri. The Society for Freshwater Science. <http://www.jnabs.org/doi/abs/10.2307/1468104>. Accessed February 2013.

Rehn, A.C., P.R. Ode, and C.P. Hawkins. 2007. Comparisons of targeted-riffle and reach-wide benthic macroinvertebrate samples—implications for data sharing in stream condition assessments. *Journal of the North American Benthological Society* 26:332–348.

Resh, V.H., and D.M. Rosenberg. 1984. *The Ecology of Aquatic Insects*. Praeger, New York.

Resh, V.H., and D.M. Rosenberg. 1989. Spatio-temporal variability and the study of aquatic insects. *Canadian Entomologist* 121:941–963.

Reynoldson, T.B., R.H. Norris, V.H. Resh, K.E. Day, and D.M. Rosenberg. 1997. The reference condition: A comparison of multimetric and multivariate approaches to assess water-quality impairment using benthic macroinvertebrates. *Journal of the North American Benthological Society* 16:833–852.

Richards, C., R.J. Haro, L.B. Johnson, and G.E. Host. 1997. Catchment and reach-scale properties as indicators of macroinvertebrate species traits. *Freshwater Biology* 37:219–230.

Ripley, J., L. Iwanowicz, V. Blazer, and C. Foran. 2008. Utilization of protein expression profiles as indicators of environmental impairment of small mouth bass (*Micropterusdolomieu*) from the Shenandoah River, Virginia, USA. *Environmental Toxicology and Chemistry* 27(8):1756–1767.

Riva-Murray, K., R.W. Bode, and P.J. Phillips. 2002. Impact source determination with biomonitoring data in New York State: Concordance with environmental data. *Northeastern Naturalist* 9(2):127–162.

Shipley, B. 2000. *Cause and Correlation in Biology: A User's Guide to Path Analysis, Structural Equations, and Causal Inference*. Cambridge University Press, Cambridge, UK.

Simpson, J.C., and R.H. Norris. 2000. Biological assessment of river quality: Development of AUSRIVAS models and outputs. Pp 125–142 in *Assessing the Biological Quality of Fresh Waters: RIVPACS and Other Techniques*.

Smucker, N.J., and M.L. Vis. 2009. Use of diatoms to assess agricultural and coal mining impacts on streams and a multiassemblage case study. *Journal of the North American Benthological Society* 28(3):659–675.

Snook, H., S.P. Davies, J. Gerritsen, B.K. Jessup, R. Langdon, D. Neils, and E. Pizutto. 2007. *The New England Wadeable Stream Survey (NEWS): Development of Common Assessments in the Framework of the Biological Condition Gradient*. Prepared for U.S. EPA Office of Science and Technology and U.S. EPA Office of Wetlands, Oceans, and Watersheds, Washington, DC by Tetra Tech, Inc., Owings Mills, MD. <http://www.epa.gov/region1/lab/pdfs/NEWSfinalReport_August2007.pdf>. Accessed November 2012.

Southerland, M., J. Vølstad, L. Erb, E. Weber, and G. Rogers. 2006. *Proof of Concept for Integrating Bioassessment Results from Three State Probabilistic Monitoring Programs*. EPA/903/R-05/003. U.S. Environmental Protection Agency, Region 3, Office of Environmental Information and Mid-Atlantic Integrated Assessment Program, Ft. Meade, MD.

Southwood, T.R.E. 1977. Habitat, the templet for ecological strategies? *Journal of Animal Ecology* 46:337–365.

Southwood, T.R.E. 1988. Tactics, strategies and templates. *Oikos* 52:3–18.

Stark, J.D. 1993. Performance of the macroinvertebrate community index: Effects of sampling method, sample replication, water depth, current velocity, and substratum on index values. *New Zealand Journal of Marine and Freshwater Research* 27(4):463–478.

Statzner, B., P. Bady, S. Dolédec, and F. Schöll. 2005. Invertebrate traits for the biomonitoring of large European rivers: An initial assessment of trait patterns in least impacted river reaches. *Freshwater Biology* 50:2136–2161.

Stribling, J.B., K.L. Pavlik, S.M. Holdsworth, and E.W. Leppo. 2008. Data quality, performance, and uncertainty in taxonomic identification for biological assessments. *Journal of the North American Benthological Society* 27(4):906–919.

Stoddard, J., D.P. Larsen, C.P Hawkins, R.K. Johnson, and R.H. Norris. 2006. Setting expectations for the ecological condition of streams: The concept of reference condition. *Ecological Applications* 16:1267–1276.

Suter, G.W., II, S.B. Norton, and S.M. Cormier. 2002. A methodology for inferring the causes of observed impairments in aquatic ecosystems. *Environmental Toxicology and Chemistry* 21(6):1101–1111.

Thienemann, A. 1954. Ein drittes biozonotisches Grundprinzip. *Archives fur Hydrobiologie* 49:421–422.

Thompson, S. 1992. *Sampling*. John Wiley & Sons, New York.

Townsend, C.R., S. Dolédec, and M.R. Scarsbrook. 1997. Species traits in relation to temporal and spatial heterogeneity in streams: A test of habitat templet theory. *Freshwater Biology* 37:367–387.

USEPA (U.S. Environmental Protection Agency). 1990. *Biological Criteria: National Program for Surface Waters*. EPA 440-5-90-004. U.S. Environmental Protection Agency, Office of Water, Washington, DC. <http://www.epa.gov/bioindicators/pdf/EPA-440-5-90-004Biologicalcriterianationalprogramguidanceforsurfacewaters.pdf>. Accessed October 2012.

USEPA (U.S. Environmental Protection Agency). 1995. *Generic Quality Assurance Project Plan Guidance for Programs Using Community Level Biological Assessment in Wadeable Streams and Rivers*. EPA 841-B-95-004. U.S. Environmental Protection Agency, Office of Water, Washington, DC.

USEPA (U.S. Environmental Protection Agency). 1998. *Lakes and Reservoir Bioassessment and Biocriteria Technical Guidance Document.* EPA 841-B-98-007. U.S. Environmental Protection Agency, Office of Water, Washington, DC. <http://www.epa.gov/owow/monitoring/tech/lakes.html>. Accessed October 2012.

USEPA (U.S. Environmental Protection Agency). 2000. *Stressor Identification Guidance Document.* EPA-822-B-00-025. U.S. Environmental Protection Agency, Office of Water and Office of Research and Development. <http://permanent.access.gpo.gov/websites/epagov/www.epa.gov/ost/biocriteria/stressors/stressorid.pdf>. Accessed February 2013.

USEPA (U.S. Environmental Protection Agency). 2001. *Biological Criteria: Technical Guidance for Streams and Small Rivers.* EPA 822-B-96-001. U.S. Environmental Protection Agency, Office of Science and Technology. <http://www.epa.gov/bioindicators/pdf/EPA-822-B-96-001BiologicalCriteria-TechnicalGuidanceforStreamsandSmallRivers-revisededition1996.pdf>. Accessed October 2012.

USEPA (U.S. Environmental Protection Agency). 2002a. *Consolidated Assessment and Listing Methodology–Toward a Compendium of Best Practices.* U.S. Environmental Protection Agency, Office of Wetlands, Oceans, and Watersheds, Washington, DC. <http://water.epa.gov/type/watersheds/monitoring/calm.cfm>. Accessed August 2012.

USEPA (U.S. Environmental Protection Agency). 2002b. *Summary of Biological Assessment Programs and Biocriteria Development for States, Tribes, Territories, and Interstate Commissions: Streams and Wadeable Rivers.* EPA-822-R-02-048. U.S. Environmental Protection Agency, Office of Environmental Information and Office of Water, Washington, DC.

USEPA (U.S. Environmental Protection Agency). 2006. *Wadeable Streams Assessment: A Collaborative Survey of the Nation's Streams.* EPA-841-B-06-002. U.S. Environmental Protection Agency, Office of Research and Development and Office of Water. <http://water.epa.gov/type/rsl/monitoring/streamsurvey/upload/2007_5_16_streamsurvey_WSA_Assessment_May2007.pdf>. Accessed January 2013.

USEPA (U.S. Environmental Protection Agency). 2010a. *Causal Analysis/Diagnosis Decision Information System (CADDIS).* U.S. Environmental Protection Agency, Office of Research and Development, Washington, DC. <http://www.epa.gov/caddis>. Last updated September 23, 2010.

USEPA (U.S. Environmental Protection Agency). 2010b. *Region V State Biological Assessment Programs Review: Critical Technical Elements Evaluation and Program Evaluation Update (2002–2010).* U.S. Environmental Protection Agency, Region V, Chicago, IL.

USEPA (U.S. Environmental Protection Agency). 2010c. *Using Stressor-response Relationships to Derive Numeric Nutrient Criteria.* EPA-820-2-10-001. U.S. Environmental Protection Agency, Office of Water, Washington, DC.

USEPA (U.S. Environmental Protection Agency). 2011a. *Aquatic Resource Monitoring – Terminology.* U.S. Environmental Protection Agency, Washington, DC. <http://www.epa.gov/nheerl/arm/terms.htm>. Accessed September 2012.

USEPA (U.S. Environmental Protection Agency). 2011b. *Biological Assessments: Key Terms and Concepts.* EPA/820/F-11/006. U.S. Environmental Protection Agency, Washington, DC. <http://water.epa.gov/scitech/swguidance/standards/criteria/aqlife/biocriteria/upload/primer_factsheet.pdf>. Accessed June 2012.

USEPA (U.S. Environmental Protection Agency). 2011c. *Primer on Using Biological Assessments to Support Water Quality Management.* EPA 810-R-11-01. U.S. Environmental Protection Agency, Washington, DC. <http://water.epa.gov/scitech/swguidance/standards/criteria/aqlife/biocriteria/upload/primer_update.pdf>. Accessed June 2012.

Van Kleef, H.H., W.C.E.P. Verberk, R.S.E.W. Leuven, H. Esselink, G. Van der Velde, and G.A. Van Duinen. 2006. Biological traits successfully predict the effects of restoration management on macroinvertebrates in shallow softwater lakes. *Hydrobiologia* 565:201–216.

Van Sickle, J., and R.M. Hughes. 2000. Classification strengths of ecoregions, catchments, and geographic clusters for aquatic vertebrates in Oregon. *Journal of the North American Benthological Society* 19:370-384.

Vannote, R.L., and B.W. Sweeney. 1980. Geographic analysis of thermal equilibria: A conceptual model for evaluating the effect of natural and modified thermal regimes on aquatic insect communities. *The American Naturalist* 115(5):667–695.

Vannote, R.L., G.W. Minshall, K.W. Cummins, J.R. Sedell, and C.E. Cushing. 1980. The River Continuum Concept. *Canadian Journal of Fisheries and Aquatic Sciences* 37:130–137.

Waite, I.R., A.T. Herlihy, D.P. Larsen, N.S. Urquart, and D.J. Klemm. 2004. The effects of macroinvertebrate taxonomic resolution in large landscape bioassessments: An example from the Mid-Atlantic Highlands, U.S.A. *Freshwater Biology* 49:474–489.

Whittier, T.R., R.M. Hughes, J.L. Stoddard, G.A. Lomnicky, D.V. Peck, and A.T. Herlihy. 2007a. A structured approach for developing indices of biotic integrity—Three examples from western streams and rivers in the USA. *Transactions of the American Fisheries Society* 136:718–735.

Whittier, T.R., R.M. Hughes, G.A. Lomnicky, and D.V. Peck. 2007b. Fish and amphibian tolerance values and an assemblage tolerance index for streams and rivers in the western USA. *Transactions of the American Fisheries Society* 136:254–271.

Yoder, C.O., and M.T. Barbour. 2009. Critical elements of state bioassessment programs: A process to evaluate program rigor and comparability. *Environmental Monitoring and Assessment* 150(1):31–42.

Yoder, C.O., and J.E. DeShon. 2003. Using biological response signatures within a framework of multiple indicators to assess and diagnose causes and sources of impairments to aquatic assemblages in selected Ohio rivers and streams. In *Biological Response Signatures: Indicator Patterns Using Aquatic Communities*, ed. T. P. Simon, pp. 23–81. CRC Press, Boca Raton, FL.

Yoder, C.O., and E.T. Rankin. 1995a. Biological criteria program development and implementation in Ohio. In *Biological Assessment and Criteria: Tools for Water Resource Planning and Decision Making*, ed. W. Davis and T. Simon, pp. 109–144. Lewis Publishers, Boca Raton, FL.

Yoder, C.O., and E.T. Rankin. 1995b. Biological response signatures and the area of degradation value: New tools for interpreting multimetric data. In *Biological Assessment and Criteria: Tools for Water Resource Planning and Decision Making*, ed. W. Davis and T. Simon, pp. 263–286. Lewis Publishers, Boca Raton, FL.

Yoder, C.O., R.J. Miltner, V.L. Gordon, E.T. Rankin, N.B. Kale, and D.K. Hokanson. 2011. *Improving Water Quality Standards and Assessment Approaches for the Upper Mississippi River: UMR Clean Water Act Biological Assessment Implementation Guidance*. Upper Mississippi River Basin Association, St. Paul, MN.

Yuan, L.L. 2010. Estimating the effects of excess nutrients on stream invertebrates from observational data. *Ecological Applications* 20(1):110–125.

Yuan, L.L., and S.B. Norton. 2003. Comparing responses of macroinvertebrate metrics to increasing stress. *Journal of the North American Benthological Society* 22(2):308–322.

Acronyms and Abbreviations

BCG	biological condition gradient
BMP	best management practice
BPJ	best professional judgment
CALM	Consolidated Assessment and Listing Methodology
CWA	Clean Water Act
DELT	deformities, erosions, lesions, and tumors
DQO	data quality objective
EOLP	Erie Ontario Lake Plain
EPA	U.S. Environmental Protection Agency
EPT	ephemeroptera, plecoptera, trichoptera taxa
FTE	full-time employee
GIS	geographic information system
HELP	Lake Huron/Lake Erie Plain
HUC	hydrologic unit code
IBI	index of biological/biotic integrity
IT	information technology
MDEP	Maine Department of Environmental Protection
MQO	measurement quality objective
NPDES	National Pollutant Discharge Elimination System
NYSDEC	New York State Department of Environmental Conservation
Ohio EPA	Ohio Environmental Protection Agency
QA	quality assurance
QC	quality control
PAH	polycyclic aromatic hydrocarbon

RDBMS   relational database management system

SOP   standard operating procedure

TMDL   total maximum daily load

UAA   use attainability analysis

WET   whole effluent toxicity

WQS   water quality standards

WSA   Wadeable streams assessment

GLOSSARY

aquatic assemblage	An association of interacting populations of organisms in a water body; for example, fish assemblage or a benthic macroinvertebrate assemblage.
aquatic community	An association of interacting assemblages in a water body, the biotic component of an ecosystem.
aquatic life use	A beneficial use designation in which the water body provides, for example, suitable habitat for survival and reproduction of desirable fish, shellfish, and other aquatic organisms.
attribute	The measurable part or process of a biological system.
benthic macroinvertebrates or benthos	Animals without backbones, living in or on the sediments, of a size large enough to be seen by the unaided eye and which can be retained by a U.S. Standard no. 30 sieve (28 meshes per inch, 0.595-mm openings); also referred to as benthos, infauna, or macrobenthos.
best management practice	An engineered structure or management activity, or combination of those, that eliminates or reduces an adverse environmental effect of a pollutant.
biological assessment or bioassessment	An evaluation of the biological condition of a water body using surveys of the structure and function of a community of resident biota.
biological criteria or biocriteria	Narrative expressions or numeric values of the biological characteristics of aquatic communities based on appropriate reference conditions; as such, biological criteria serve as an index of aquatic community health.
biological indicator or bioindicator	An organism, species, assemblage, or community characteristic of a particular habitat, or indicative of a particular set of environmental conditions.
biological integrity	The ability of an aquatic ecosystem to support and maintain a balanced, adaptive community of organisms having a species composition, diversity, and functional organization comparable to that of natural habitats in a region.

biological monitoring or biomonitoring	Use of a biological entity as a detector and its response as a measure to determine environmental conditions; ambient biological surveys and toxicity tests are common biological monitoring methods.
biological survey or biosurvey	Collecting, processing, and analyzing a representative portion of the resident aquatic community to determine its structural and/or functional characteristics.
Clean Water Act	The act passed by the U.S. Congress to control water pollution (formally referred to as the Federal Water Pollution Control Act of 1972). Public Law 92-500, as amended. 33 U.S.C. 1251 *et seq.*
Clean Water Act 303(d)	This section of the act requires states, territories, and authorized tribes to develop lists of impaired waters for which applicable WQS are not being met, even after point sources of pollution have installed the minimum required levels of pollution control technology. The law requires that the jurisdictions establish priority rankings for waters on the lists and develop TMDLs for the waters. States, territories, and authorized tribes are to submit their lists of waters on April 1 in every even-numbered year.
Clean Water Act 305(b)	Biennial reporting requires description of the quality of the nation's surface waters, evaluation of progress made in maintaining and restoring water quality, and description of the extent of remaining problems.
criteria	Elements of state water quality standards, expressed as constituent concentrations, levels, or narrative statements, representing a quality of water that supports a particular use. When criteria are met, water quality will generally protect the designated use.
DELT	Presence of deformities, erosions, lesions, and tumors as a measure of organism health, typically assessed for fish.
designated uses	Those uses specified in WQS for each water body or segment whether or not they are being attained.
disturbance	Human activity that alters the natural state and can occur at or across many spatial and temporal scales.
ecoregion	A relatively homogeneous ecological area defined by similarity of climate, landform, soil, potential natural vegetation, hydrology, or other ecologically relevant variables.

function	Processes required for normal performance of a biological system (might be applied to any level of biological organization).
guild	A group of organisms that exhibit similar habitat requirements and that respond in a similar way to changes in their environment.
historical data	Data sets from previous studies, which can range from handwritten field notes to published journal articles.
index of biological/biotic integrity	An integrative expression of site condition across multiple metrics; an IBI is often composed of at least seven metrics.
invasive species	A species whose presence in the environment causes economic or environmental harm or harm to human health. Native species or nonnative species can show invasive traits, although that is rare for native species and relatively common for nonnative species. (Note that this term is not included in the biological condition gradient [BCG].)
least disturbed condition	The best available existing conditions with regard to physical, chemical, and biological characteristics or attributes of a water body within a class or region. Such waters have the least amount of human disturbance in comparison to others in the water body class, region, or basin. Least disturbed conditions can be readily found but can depart significantly from natural, undisturbed conditions or minimally disturbed conditions. Least disturbed condition can change significantly over time as human disturbances change.
metric	A calculated term or enumeration that represents some aspect of biological assemblage, function, or other measurable aspect and is a characteristic of the biota that changes in some predictable way with increased human influence.
minimally disturbed condition	The physical, chemical, and biological conditions of a water body with very limited, or minimal, human disturbance.
multimetric index	An index that combines indicators, or metrics, into a single index value. Each metric is tested and calibrated to a scale and transformed into a unitless score before being aggregated into a multimetric index. Both the index and metrics are useful in assessing and diagnosing ecological condition. See **index of biological/biotic integrity.**

narrative biological criteria	Written statements describing the structure and function of aquatic communities in a water body that support a designated aquatic life use.
native	An original or indigenous inhabitant of a region; naturally present.
nonnative or intentionally introduced species	With respect to an ecosystem, any species that is not found in that ecosystem; species introduced or spread from one region of the United States to another outside their normal range are nonnative or non-indigenous, as are species introduced from other continents.
numeric biological criteria	Specific quantitative measures of the structure and function of aquatic communities in a water body necessary to protect a designated aquatic life use.
periphyton	A broad organismal assemblage composed of attached algae, bacteria, their secretions, associated detritus, and various species of microinvertebrates.
rapid bioassessment protocols	Cost-effective techniques used to survey and evaluate the aquatic community to detect aquatic life impairments and their relative severity.
reference condition (biological integrity)	The condition that approximates natural, unaffected conditions (biological, chemical, physical, and such) for a water body. Reference condition (biological integrity) is best determined by collecting measurements at a number of sites in a similar water body class or region undisturbed by human activity, if they exist. Because undisturbed conditions can be difficult or impossible to find, minimally or least disturbed conditions, combined with historical information, models, or other methods can be used to approximate reference condition as long as the departure from natural or ideal is understood. Reference condition is used as a benchmark to determine how much other water bodies depart from this condition because of human disturbance. See **minimally disturbed condition** and **least disturbed condition**

reference site

A site selected for comparison with sites being assessed. The type of site selected and the types of comparative measures used will vary with the purpose of the comparisons. For the purposes of assessing the ecological condition of sites, a reference site is a specific locality on a water body that is undisturbed or minimally disturbed and is representative of the expected ecological integrity of other localities on the same water body or nearby water bodies.

sensitive taxa

Taxa intolerant to a given anthropogenic stress; first species affected by the specific stressor to which they are *sensitive* and the last to recover following restoration.

sensitive or regionally endemic taxa

Taxa with restricted, geographically isolated distribution patterns (occurring only in a locale as opposed to a region), often because of unique life history requirements. Can be long-lived, late-maturing, low-fecundity, limited-mobility, or require mutualist relation with other species. Can be among listed endangered/threatened or special concern species. Predictability of occurrence often low; therefore, requires documented observation. Recorded occurrence can be highly dependent on sample methods, site selection, and level of effort.

sensitive - rare taxa

Taxa that naturally occur in low numbers relative to total population density but can make up large relative proportion of richness. Can be ubiquitous in occurrence or can be restricted to certain microhabitats, but because of low density, recorded occurrence is dependent on sample effort. Often stenothermic (having a narrow range of thermal tolerance) or coldwater obligates; commonly k-strategists (populations maintained at a fairly constant level; slower development; longer lifespan). Can have specialized food resource needs or feeding strategies. Generally intolerant to significant alteration of the physical or chemical environment; are often the first taxa observed to be lost from a community.

sensitive - ubiquitous taxa

Taxa ordinarily common and abundant in natural communities when conventional sample methods are used. Often having a broader range of thermal tolerance than sensitive or rare taxa. These are taxa that constitute a substantial portion of natural communities and that often exhibit negative response (loss of population, richness) at mild pollution loads or habitat alteration.

stressors Physical, chemical, and biological factors that adversely affect aquatic organisms.

structure Taxonomic and quantitative attributes of an assemblage or community, including species richness and relative abundance structurally and functionally redundant attributes of the system and characteristics, qualities, or processes that are represented or performed by more than one entity in a biological system.

taxa A grouping of organisms given a formal taxonomic name such as species, genus, family, and the like.

taxa of intermediate tolerance Taxa that compose a substantial portion of natural communities; can be r-strategists (early colonizers with rapid turnover times; boom/bust population characteristics). Can be eurythermal (having a broad thermal tolerance range). Can have generalist or facultative feeding strategies enabling use of relatively more diversified food types. Readily collected with conventional sample methods. Can increase in number in waters with moderately increased organic resources and reduced competition but are intolerant of excessive pollution loads or habitat alteration.

threatened waters Waters that are currently attaining water quality standards, but which are expected to exceed water quality standards by the next 303(d) listing cycle.

tolerant taxa Taxa that compose a small proportion of natural communities. They are often tolerant of a broader range of environmental conditions and are thus resistant to a variety of pollution- or habitat-induced stresses. They can increase in number (sometimes greatly) in the absence of competition. Commonly r-strategists (early colonizers with rapid turnover times; boom/bust population characteristics), able to capitalize when stress conditions occur; last survivors.

total maximum daily load The calculated maximum amount of a pollutant a water body can receive and still meet WQS and an allocation of that amount to the pollutant's source.

water quality management (nonregulatory) Decisions on management activities relevant to a water resource, such as problem identification, need for and placement of best management practices, pollution abatement actions, and effectiveness of program activity.

water quality standard

A law or regulation that consists of the designated use or uses of a water body, the narrative or numerical water quality criteria (including any biological criteria) that are necessary to protect the use or uses of that water body, and an antidegradation policy.

whole effluent toxicity

The aggregate toxic effect of an aqueous sample (e.g., whole effluent wastewater discharge) as measured by an organism's response after exposure to the sample (e.g., lethality, impaired growth or reproduction); WET tests replicate the total effect and actual environmental exposure of aquatic life to toxic pollutants in an effluent without requiring the identification of the specific pollutants.

APPENDIX A: AGENDA FOR ON-SITE INTERACTION MEETING

State/Tribal Agency Biological Assessment Program Evaluation

AGENDA

<u>DAY 1</u>   <u>Date</u>

                  Building #_____ Room _____

9:30–10:00 am   **Welcome and Introductions**

- Refinements to the agenda

- General purpose and overview

10:00–11:30   **[Agency] Biological Assessment Program Review & Development**

- Key concepts and examples

- Development of state programs

- U.S. Environmental Protection Agency (EPA) methods and key documentation

11:30–1:00 pm   **LUNCH**

1:00–2:00   **Overview of [name of water quality agency to be reviewed] Biological Assessment Program by [Agency] staff**

- Brief history of [water quality agency] biological program

- Current developments and updates

2:00–5:00   **[Agency] Monitoring & Assessment Program—following list of annotated discussion topics**

Monitoring & Assessment Program

- Water body types

- Spatial design

- Basin assessments

- Indicators—chemical, physical, biological

- Data management

- Resources for monitoring and assessment

Reporting & Listing

- Delineation of impairments

- Assessment process

- 305(b)/303(d)

- Other program support

DAY 2   Date

Building #_____ Room _____

9:00–10:30 am   **[Agency] Managers' Overview of Biological Assessment-based Programs**

- Process overview

- Concepts and examples–implications for water quality standards (WQS)

10:30–11:30   **Assessment and Integration**

- Using indicators to measure effectiveness

- Using monitoring and assessment to support water quality management programs

11:30–1:00 pm   **LUNCH**

1:00–3:00   **Water Quality Standards**

- General description of [Agency] WQS

- Structure of designated uses and attendant criteria

- Aquatic life uses and biological criteria

- Use attainability analyses (UAAs), site-specific modifications, etc.

- Implications

3:00–5:00   **Agency Self-Assessments**

- Complete agency self-assessments and discuss results (might be beneficial to have the agency complete the self-assessments prior to the biological assessment program evaluation)

<u>DAY 3</u>   <u>Date</u>

Building #_____ Room _____

8:30–11:30 am   **Technical Elements Review of [Agency] Biological assessment Program**

- Overview of technical elements review process

- Scoring each element in the technical elements checklist

11:30–1:00 pm   **LUNCH**

1:00–2:00   **Technical Elements Review (continued)**

2:00–4:30   **Q&A**

- Follow-up on any of the previous days' topics

APPENDIX B: INTERVIEW TOPICS FOR AGENCY REVIEW

State/Tribal Monitoring and Assessment and Water Quality Standards Program Interviews:
Annotated List of Discussion Topics

Introduction

A critical component of the biological program review is the detailed interviews of key agency program managers and staff. The purpose of these discussions is to understand the existence and extent of data-driven water quality management. These interviews are an opportunity to better define and understand the uses of monitoring and assessment information in the water quality agency and to determine the opportunities, incentives, impediments, and barriers to the fuller use of this information in support of water quality management programs. In addition, the interviews examine the intersections of biological assessment with water quality standards (WQS), designated aquatic life uses, and criteria.

The biological program review is focused on current and planned uses of monitoring and assessment information in support of all relevant water quality management programs. This includes the following broad program areas that water quality management agencies have in common:

- WQS focusing on designated uses and criteria
- Reporting and listing (watershed assessments, Clean Water Act [CWA] section 305(b)/303(d) reporting) and total maximum daily load (TMDL) development schedules
- Water quality planning, TMDL development and implementation, nonpoint source assessment and management, dredge and fill (CWA section 404/401)
- National Pollutant Discharge Elimination System (NPDES) program (CWA section 402)

Managers and staff who can speak to the operation and management of these programs should attend the interview when these topics are discussed.

The following topics are intended to guide the interview process. These topics are also intended to help the agency determine who from the agency programs should attend each day's discussions.

Monitoring and Assessment Program

Monitoring is the systematic collection of chemical, physical, and biological (including WET) data in the ambient environment. Assessment is the analysis and transformation of that data into meaningful information that includes attainment/nonattainment determinations, characterization of impairments (extent and severity), associations between impaired status and causes (i.e., agents) and sources (i.e., activity or origin), and data and information to develop improved tools, indicators, criteria, and policies. Monitoring and assessment supports the reporting that is required by the CWA (sections 305[b], 303[d] list, 319, etc.) and that is used by the agency for allied purposes (watershed assessments, site-specific assessments, planning, TMDL development, etc.). The following are core topics for discussion. The agency might wish to add other topics.

1. Spatial design

    • Is a rotating basin approach used? Describe the sequence and cycle and, linkages to management activities.

    • Is the spatial design probability-based (scale and scope, statewide, regional, etc.)?

    • Fixed station (e.g., tenure and history)

    • What resource types are covered (wadeable streams, large rivers, great rivers, lakes, wetlands, headwater streams, etc.)?

    • Is the spatial design for the monitoring program aligned with, or directly feeding into, other monitoring and assessment programs at the local, regional, or federal level?

2. Basin assessments

    • At what scale are assessments done (major basin, subbasin, watershed, subwatershed)? Hydrologic unit code (HUC) units?

    • What is the site-selection process (targeted, random, other)?

    • What stratifying factors are considered (watershed area, stream order, other)?

    • How many sites are assessed each year?

    • What site density (i.e., the number of sites allocated to a specific study area) is used?

    • What is the data analysis and reporting sequence?

    • What are the bottlenecks in data analysis and reporting?

    • Are there other significant logistical issues?

    • What study planning process is used? Are all affected disciplines integrated?

3. Index periods

    • Describe the seasonal sampling index periods by indicator (summer-fall, monthly, other).

    • Explain the flow attenuated considerations (loading estimates, event related, summer-fall low flow, etc.).

4. Biological (including WET)/chemical/physical assessment

    • What media are assessed (water, sediment, tissues, etc.)?

- What is the purpose of sampling (ambient characterization, model calibration, long-term trends, reference/background, etc.)?

- Which parameter groups are considered? How are the groups selected?

- What type of laboratory support is available?

- Describe the sampling design and logistics (survey design, frequency, grabs vs. composites).

- Are there exceedance issues (magnitude, duration, frequency)?

5. Reference condition

- Have reference sites been established? For what purposes (e.g., biological criteria, nutrients, background conditions)?

- How many reference sites are used?

- What is the spatial organization and stratification (ecoregions, hydrologic units, physiographic regions, other)?

- How is reference condition established (data driven, cultural, least affected)?

6. Data processing and management

- How are data stored (WQX, other system)?

- How are data accessed by staff for analysis?

- What resources are dedicated to data management (full time employees [FTEs])?

- What are the quality assurance/quality control (QA/QC) procedures for ensuring data quality?

- What is the timetable for entry and validation?

- Describe the ease of data availability within and outside the agency.

- What is the demand for data from outside the agency?

7. Monitoring strategy

- Discuss the latest monitoring strategy available (please provide a copy).

- Is the strategy a useful document?

- Should the strategy serve as documentation of data acceptability?

- Are data quality objectives (DQOs) defined?

- How frequently is the strategy updated?

8. Resources

- How many FTEs are devoted to monitoring and assessment by discipline (chemical/physical, biological assessment, TMDL/modeling, etc.)?

- What proportion of FTEs is devoted to water quality management programs? (provide a table of organization for the CWA parts of the water quality agency program)

- What funding sources are available? What are their limitations? Is the agency leveraging resources with other programs?

- Are current resources adequate? If not, what is needed?

Reporting and Listing (305[b]/303[d]) and TMDLs

Reporting and listing are the processes of producing the integrated CWA section 305(b)/303(d) report, which includes the list of waters with impaired or threatened uses and TMDL development schedules. The information contained in these reports and lists is not only important to determining the effectiveness of a water quality agency's water quality management programs, but is increasingly being used to set program priorities and allocate funding. Monitoring and assessment information is an indispensable element of this process and how it is generated and applied determines, in part, the accuracy of the statistics that are reported and used. Thus, it is important to determine and understand how each water quality agency uses monitoring and assessment information to support these determinations.

1. Delineation of impaired or threatened waters

- What are the procedures and protocols for determining impaired waters (including extent and severity)?

- What are the primary arbiters of impairment and threat?

- What data qualifiers are used (analogs to the formerly used monitored and evaluated categories)?

- What is the extent of extrapolation from single and aggregate sampling sites? How was this developed, and has it been tested?

- What data are the basis of decisions about aquatic life use impairment (biological, chemical/physical, mix of both, best professional judgment [BPJ], etc.)?

- Is determination of causes and sources of impairment and threat linked to an impairment or threat?

- How are determinations of severity, extent, and incremental change made?

- How is the universe of resources defined (miles of rivers and streams, lake acres, etc.)?
- How does the water quality agency account for the proportion of resources that are actually assessed?

2. Assessment process

- Explain "chain-of-custody." Do the same staff who collect and analyze sampling data also produce the assessments? Are there any "hand-offs"?
- How are data from volunteer organizations used? Are there "admission" requirements? Any testing of accuracy? Pressure to accept data?
- How are data from other organizations handled? What are the acceptance requirements?
- Are there requirements for credible data or similar legislation?

3. 305(b) reporting topics

- How are trends assessed (e.g., tracking of aggregate condition through time, by resource type, designated uses, etc.)?
- How is CWA section 305(b) reporting information used by agency to guide water quality management? Is it the 305(b) report viewed by management as a report card? Does it have other uses? Does it distinguish impairment by point and nonpoint sources? Any subsets within each?
- What is the extent to which outside groups use 305(b) reporting information?
- What would be the impact of any changes due to assessment method?

4. 303(d) listing and TMDLs

- Describe the relationship between former CWA section 305(b) report and existing 303(d) list (e.g., conversion process, issues, concerns, gaps, and shortfalls).
- Is TMDL development coordinated or aligned with ambient monitoring and assessment?
- Are biological data used in the TMDL process? Are there any issues and concerns? Conflicts?
- How are biological impairments considered? Which listing category?
- Are there sufficient biological assessment tools available to help develop defensible TMDLs that will contribute to restoration of impaired aquatic life uses? If not, what is needed and how long will it take?

Water Quality Standards

WQS provide the basis for water quality management and for judging the effectiveness of water quality management programs.

- General WQS issues

  - Describe the structure of the water quality agency's current WQS (designated uses, criteria, and antidegradation policy and implementation procedures).

  - How are chemical water quality criteria derived? Any modifiers or adjustment factors?

  - How are existing uses determined?

  - When and where are site-specific criteria used? How many instances?

  - How would better monitoring and assessment affect the WQS process?

- Designated uses

  - Describe aquatic life designated uses in the state WQS (a copy of the relevant parts of the WQS is requested).

  - Are individual waters designated? Are there default uses? Undesignated waters? Tributary rules? Other issues?

  - What triggers individual water body designations? Are they always downgrades? Does anything trigger an upgrade? Is there a regular process for inventorying these needs?

  - Are there designated uses that are less than the CWA section 101(a)(2) goal uses? Are they defined?

  - Is there a process to use biological assessments to more precisely define designated aquatic life uses and develop numeric biological criteria to protect those uses?

  - What is the level of water quality agency interest in use of biological assessment to more precisely define uses (advantages, disadvantages, barriers to development and implementation)?

- Use attainability analysis (UAA)

  - Does the agency have experience with UAAs (number attempted/completed, problems, issues)?

  - Outline/describe the existing UAA process. Is it routine? Special project oriented? What triggers a UAA? What are preferred data and information requirements?

- How do stakeholders perceive the UAA process (pros and cons, requests for and by whom, etc.)?

- Has the emphasis on CWA section 303(d) listing increased the "interest" in UAAs?

- What criteria are used to determine attainability of uses?

- What are the likely stressors in your state? What are the sources of the stressors?

- Biological criteria

  - Have biological criteria been adopted or proposed (narrative, numeric)?

  - How are biological criteria linked to designated uses?

  - Are biological assessments used to more precisely define designated aquatic life uses and develop numeric biological criteria?

  - What are the advantages and disadvantages of biological criteria in WQS?

  - How would numeric biological criteria affect the use review process?

  - Describe habitat assessments and criteria.

  - What are stakeholder perceptions and viewpoints on biological criteria?

Assessment Integration Issues

The integration of monitoring and assessment information within water quality management programs is an important and emerging issue. The National Environmental Performance Partnership System promotes joint priority setting and planning through the increased use of environmental goals and indicators. Shared goals and milestones could be used to more comprehensively report to the public and environmental decision makers about the status of water resources in the water quality agency and to document progress in meeting these goals. The goals and milestones could also be used to more effectively target programmatic efforts at all levels. It is important to be able document achievements so that environmental successes are recognized, funding is maintained at appropriate levels, and effective management programs continue to be implemented. The following are aimed at assessing the water quality agency's efforts to develop and use indicators and integrate them into water quality management.

1. Indicators for surface waters

   - What efforts have been taken to develop a process for using environmental indicators to fulfill the role as a measure of the effectiveness of water quality management programs (provide any documentation)?

   - Are any implemented or practiced?

   - How dependent are these systems on monitoring data?

- What is the awareness of past U.S. Environmental Protection Agency (EPA) indicator development efforts (i.e., national indicators for surface waters, hierarchy of indicators, etc.)?

- Is there any recognition of indicator roles (i.e., stress, exposure, response roles of indicators)?

- What is (are) the most important measure(s) or indicator(s) of water quality management program success in your water quality agency?

2. Program integration

- Are there any examples in which water quality management programs rely on ambient monitoring and assessment information?

- Is monitoring and assessment information used to support:

  o The NPDES permitting process (e.g., reasonable potential determinations and permit compliance)? CWA section 402 NPDES program including stormwater phase I or II?

  o CWA section 319/nonpoint source planning and implementation?

  o CWA 404/401 process? Other programs?

- How is monitoring and assessment information and resulting assessments and reports, regarded by the above programs (essential, useful, nice to have, inconsequential)?

3. Training

- Are training opportunities afforded to staff and/or management?

- How do these relate to indicators development, monitoring and assessment, biological assessment, or ecological principles in general?

- Does your agency receive requests for field demonstrations (fish, bugs, sampling, etc.) for internal and external purposes?

- Is training available for external entities?

APPENDIX C: SELF-ASSESSMENTS BY STATE/TRIBAL AGENCY MANAGERS

The self-assessment exercise is conducted during the on-site evaluation. The technical expert walks participants through a discussion of how biological assessment information can be more effectively used to support water quality program needs for information. It is important that representatives from different water quality programs participate in order to: (1) gain a cross-program understanding of how biological assessments can be used to support multiple water quality programs; (2) identify the type of biological assessment information needed by their programs and timing for information delivery; and, (3) identify efficiencies for more cost effective biological assessments. Programs interested in conducting a review do not need to complete these self-assessment questions in advance. The results of these discussions do not factor into scoring of the technical elements.

The topics and questions included in the worksheets are provided as examples that can be used to initiate cross program discussion.

SELF-ASSESSMENT 1

Use of biological assessments to protect aquatic life use

1. Answering these questions requires a thorough understanding of the aquatic life uses in your water quality agency's water quality standards law.

   - To know this, you have to be familiar with the aquatic life uses in your water quality standards and understand what parts, if any, of the aquatic life uses are assessed with biological assessment data.

2. For aquatic life uses that are assessed using biological assessment data, an estimate of what biological condition gradient (BCG) level, or levels, your water quality agency's uses provide protection is recommended;

   - To know this, the biological monitoring technical staff can determine (for example, by a consensus of professional judgment) to what BCG level(s) your water quality agency's biological criteria thresholds (e.g., numeric criteria, Rapid Bioassessment Protocol (RBP), or Index of Biological Integrity (IBI) ranges) provide protection. Alternatively, if your program does not have numeric biological criteria, the staff can evaluate what BCG level your state uses for listing biologically impaired waters. In other words, how does biologically-based aquatic life use attainment measured by numeric biological criteria and/or CWA section 303(d)-listing thresholds map to a BCG level?

   - Familiarity with your water quality agency's application of biological criteria thresholds in regulatory decision-making is important to help identify how biological assessment information can be used to guide the discussion on added value of further technical improvement (i.e., be familiar with findings that have triggered an agency response based on aquatic life use attainment as determined by biological assessment and criteria).

   - Example scenarios characteristic of situations your agency encounters are recommended to help focus the discussion and the identification of current strengths and limitations of the biological assessment program.

101

WORKSHEET FOR TOPIC 1: PROTECTION OF HIGH QUALITY WATERS

Example: A watershed with minimal impacts to aquatic systems from anthropogenic stress. Streams, wetlands, lakes, and rivers support high quality biological communities based on biological indices (e.g., benthic macroinvertebrates, algal, and/or fish assemblages). The presence of reproducing native species is documented. Downstream waters such as bays and estuaries support a range of biological conditions, including high quality biological communities in areas that are minimally impacted.

1.   Does the existing biological assessment program provide information to detect declines in biological condition in high quality waters?

_____YES   _____NO

2.   If yes, does the program provide information to detect declines within the assigned aquatic life use class?

_____YES   _____NO

If no to either of the above two questions, what changes to the type, amount, or quality of biological assessment information would be useful? Would changes to data collection and analysis and/or internal communication contribute to the use of biological assessments? Are there additional recommendations?

WORKSHEET FOR TOPIC 1: PROTECTION OF HIGH QUALITY WATERS   (page 2)

3.   Does the existing biological assessment program provide information to support an agency action to assign the highest quality waters to different aquatic life use categories?

_____YES   _____NO

If no, what changes to the type, amount, or quality of biological assessment information would be useful? Would changes to data collection and analysis and/or internal communication contribute to the use of biological assessments? Are there additional recommendations?

4.   Does the existing biological assessment program currently provide information to support agency decisions and actions (e.g., antidegradation policies, best management practices) to protect the highest quality waters?

_____YES   _____NO

If no, what changes to the type, amount, or quality of biological assessment information would be useful? Would changes to data collection and analysis and/or internal communication contribute to the use of biological assessments? Are there additional recommendations?

WORKSHEET FOR TOPIC 2: PROTECTION OF CURRENT CONDITIONS

Example: A watershed with a mix of minimal to moderate impacts to aquatic systems from anthropogenic stress. Streams, wetlands, lakes, and rivers support a range of biological conditions based on biological indices (e.g., benthic macroinvertebrates, algal, and/or fish assemblages). The presence of reproducing native species has been observed in waters where there is minimal anthropogenic stress. Downstream waters such as bays and estuaries also support a comparable range of biological conditions and levels of anthropogenic stress.

1.   Does the existing biological assessment program provide information to detect declines in biological condition?

_____YES   _____NO

2.   If yes to above, are the current indices sufficiently sensitive to detect incremental declines within the assigned aquatic life use class?

_____YES   _____NO

If no to either of the above questions, what changes to the type, amount, or quality of biological assessment information would be useful? Would changes to data collection and analysis and/or internal communication contribute to the use of biological assessments? Are there additional recommendations?

WORKSHEET FOR TOPIC 2: PROTECTION OF CURRENT CONDITIONS   (page 2)

3.   Does the biological assessment program provide information that the agency could use to evaluate potential impacts on the aquatic community? (For example, a new and/or modification to an existing industrial, transportation, or residential development is proposed that might have an impact on aquatic life in the watershed.)

_____YES   _____NO

If no, what changes to the type, amount, or quality of biological assessment information would be useful? Would changes to data collection and analysis and/or internal communication contribute to the use of biological assessments? Are there additional recommendations?

WORKSHEET FOR TOPIC 2: PROTECTION OF CURRENT CONDITIONS   **(page 3)**

4.   If an evaluation for potential impacts indicates that the proposed activity would result in a further decline in biological condition, would the biological assessment information used in the evaluation support an agency action to minimize or prevent the predicted decline?

_____YES   _____NO

If yes, what changes to the type, amount, or quality of biological assessment information would be useful to provide better support?

If no, what changes to the type, amount, or quality of biological assessment information would be useful? Would changes to data collection and analysis and/or internal communication contribute to the use of biological assessments? Are there additional recommendations?

WORKSHEET FOR TOPIC 3: PROTECTION OF IMPROVED CONDITIONS

Example: A watershed with mix of minimal to severe impacts from anthropogenic stress. Streams, wetland, lakes, and rivers support a range of biological conditions from poor to excellent based on biological indices (e.g., benthic macroinvertebrates, algal, and/or fish assemblages). The presence of reproducing native species is documented only in higher quality waters. Some of the severely impacted waters have been assigned a limited or modified aquatic life use based on the findings of a use attainability analysis. Incremental improvements in biological conditions in several water bodies have been observed. For a few of the severely impacted waters, incremental improvements have been observed but conditions still do not meet a higher use class. Downstream waters such as bays and estuaries also support a comparable range of biological conditions and levels of anthropogenic stress.

1.   Does the existing biological assessment program provide information to detect incremental improvements in biological condition?

_____YES   _____NO

2.   If yes to above, are the current indices sufficiently sensitive to detect incremental changes within the assigned aquatic life use class?

_____YES   _____NO

If no to either of the two questions above, what changes to the type, amount, or quality of biological assessment information would be useful? Would changes to data collection and analysis and/or internal communication contribute to the use of biological assessments? Are there additional recommendations?

WORKSHEET FOR TOPIC 3: PROTECTION OF IMPROVED CONDITIONS   **(page 2)**

3.   Does the biological assessment program produce information to support an agency decision to report and take action to protect improved aquatic life condition in a water body where incremental improvements have been observed?

_____YES   _____NO

If yes, please identify the specific management programs currently supported by biological assessment data. Are there improvements to the type, quality, or delivery of the data that can enhance use of the data?

If no, what changes to the type, amount, or quality of biological assessment information would be useful? Would changes to data collection and analysis and/or internal communication contribute to the use of biological assessments? Are there additional recommendations?

WORKSHEET FOR TOPIC 4: SUPPORT USE CLASSIFICATION

Example: A watershed with a mix of minimal to severe impacts from anthropogenic stress. Streams, wetlands, lakes, and rivers support range of biological conditions from poor to excellent based on biological indices (e.g., benthic macroinvertebrates, algal, and/or fish assemblages). The presence of reproducing native species in the higher quality waters is well documented.

1.   Does the biological assessment program produce information to support refining an aquatic life use goal for water bodies?

_____YES   _____NO

If no, what changes to the type, amount or quality of biological assessment information would be useful? Would changes to data collection and analysis and/or internal communication contribute to the use of biological assessments? Are there additional recommendations?

SUMMARY WORKSHEET: SELF ASSESSMENT SESSION 1

Discussion Topics	YES	NO
1. Protect high quality waters		
2. Protect current conditions		
3. Protect improved conditions		
4. Support for use classification		

Summary observations and key recommendations:

SELF-ASSESSMENT 2

Use of biological assessments to support water quality management programs

1. To answer these questions requires a thorough understanding of the information flow and management decision-making process within and between programs in your agency. In some cases this communication and decision-making may primarily occur at the technical staff level, but in other cases it may occur between program managers (e.g., between the permitting and the monitoring manager, or the water quality standards coordinator and the monitoring manager) or even at the level of the water Program Director or agency Commissioner.

   - The questions are most usefully answered during a cross-program group discussion that includes representatives from all programs and levels of management.

2. For state agencies with aquatic life uses that are assessed using biological monitoring data, it is helpful to estimate to what BCG level, or levels, your water quality agency's aquatic life uses and numeric biological criteria provide protection;

   - To know this, the biological monitoring technical staff can determine (for example, by a consensus of professional judgment) to what BCG level(s) your water quality agency's biological criteria thresholds (e.g., numeric criteria, RBP, modeled index (e.g. RIVPACS), or IBI ranges) provide protection. Alternatively, if your program does not have numeric biological criteria, the staff can evaluate what BCG level your state uses for listing biologically impaired waters. In other words, how does biologically-based aquatic life use attainment measured by numeric biological criteria and/or CWA section 303(d)-listing thresholds map to a BCG level?

3. The group answering this self-assessment should have some familiarity with your water quality agency's application of biological criteria thresholds in regulatory decision-making (i.e., be familiar with findings that have triggered an agency response based on aquatic life use attainment/non-attainment as determined by biological assessment and criteria).

4. Example scenarios characteristic of situations your agency encounters are recommended to help focus the discussion and the identification of current strengths and limitations of the biological assessment program.

WORKSHEET FOR TOPIC 1: SUPPORT FOR WATER QUALITY STANDARDS

1. Does the biological assessment program provide data to support derivation of numeric biological criteria?

_____YES   _____NO

If yes, please list the water body types for which numeric biological criteria have been developed:

Primary Headwater Streams	_____YES	_____NO
Streams	_____YES	_____NO
Rivers	_____YES	_____NO
Large Rivers	_____YES	_____NO
Lakes	_____YES	_____NO
Wetlands	_____YES	_____NO
Estuaries	_____YES	_____NO
Other (add below)	_____YES	_____NO
[water body type]	_____YES	_____NO
[water body type]	_____YES	_____NO
[water body type]	_____YES	_____NO

If yes to any of the above, are there improvements or refinements to the type, amount, quality, or delivery of the data that would be useful? Please specify any recommendations for further technical development.

If no to any of the above, what changes to the type, amount, or quality of biological assessment information would be useful? Would changes to data collection and analysis and/or internal communication contribute to the use of biological assessments? Are there additional recommendations?

WORKSHEET FOR TOPIC 1: SUPPORT FOR WATER QUALITY STANDARDS   **(page 2)**

2. Does biological assessment information, whether from monitoring or from peer reviewed literature, contribute to review of existing water quality criteria and/or to detection of the need for new criteria or site-specific modifications?

_____YES   _____NO

If no, what changes to the type, amount, or quality of biological assessment information would be useful? Would changes to data collection and analysis and/or internal communication contribute to the use of biological assessments? Are there additional recommendations?

WORKSHEET FOR TOPIC 1: SUPPORT FOR WATER QUALITY STANDARDS   **(page 3)**

3. Has your agency ever used biological assessments to assess effects or determine the need for criteria for observed stressors for which there are no existing criteria?

Potential examples are listed below.

Habitat alteration	_____YES	_____NO
Water withdrawal/flow alterations	_____YES	_____NO
Suspended sediment	_____YES	_____NO
Nutrient effects	_____YES	_____NO
Other [list below if needed]	_____YES	_____NO

If no, what changes to the type, amount, or quality of biological assessment information would be useful? Would changes to data collection and analysis and/or internal communication contribute to the use of biological assessments? Are there additional recommendations?

WORKSHEET FOR TOPIC 1: SUPPORT FOR WATER QUALITY STANDARDS   **(page 4)**

4. During a triennial review, does the biological assessment program provide a list of waters that are attaining biological conditions higher than their currently assigned aquatic life use?

_____YES   _____NO

If no, what changes to the type, amount, or quality of biological assessment information would be useful? Would changes to data collection and analysis and/or internal communication contribute to the use of biological assessments? Are there additional recommendations?

WORKSHEET FOR TOPIC 1: SUPPORT FOR WATER QUALITY STANDARDS   **(page 5)**

5. Does the biological assessment program produce information to support designating a water body to an antidegradation tier?

_____YES   _____NO

If no, what changes to the type, amount, or quality of biological assessment information would be useful? Would changes to data collection and analysis and/or internal communication contribute to the use of biological assessments? Are there additional recommendations?

WORKSHEET FOR TOPIC 2: SUPPORT FOR CWA SECTION 303(D) AND TMDL PROGRAMS

1.   Does the biological assessment program provide data and information used to support assessments for CWA section 303(d) purposes?

_____YES   _____NO

If yes, what changes to the type, amount, or quality of biological assessment information and/or the timing of data availability improve support to your program? (Please provide specific recommendations.)

If no, what additional type, amount, or quality of biological assessment information would be useful? Would changes to data collection and analysis and/or internal communication contribute to the use of biological assessments? Are there additional recommendations?

WORKSHEET FOR TOPIC 2: SUPPORT FOR CWA SECTION 303(D) AND TMDL PROGRAMS (page 2)

2.   If biological assessment data has been used as the sole basis for putting one or more waters on the 303(d) list (Category 5 of the Integrated Reporting Guidance [IRG]) for failure to fully support the designated aquatic life use, was the non-support determination based on:

   2a. Failure to meet a state numeric biological criteria?   Or

   2b. Conditions inconsistent with one or more narrative WQC?

   _____YES _____NO

   If yes for 2b, was the determination regarding failure to meet narrative water quality criteria based on:

   —   Numeric biological thresholds issued as guidance values, rather than having been incorporated into the state's WQS regulations _____

   —   Qualitative guidance on how to interpret biological assessment data _____

   —   Primarily, the best professional guidance of state agency staff _____

If yes for any of these aspects, what changes to the type, amount, or quality of biological assessment information and/or the timing of data availability would improve use of biological assessments as sole basis for 303(d) listing of water bodies?

If no, what changes to the type, amount, or quality of biological assessment information might lead to use of biological assessments as the sole basis for 303(d) listing of water bodies?

WORKSHEET FOR TOPIC 2: SUPPORT FOR CWA SECTION 303(D) AND TMDL PROGRAMS (page 3)

3.   Has biological assessment data been used (in the absence of evidence of failure to meet one or more chemical or physical water quality criteria) as the basis for making an affirmative determination that one or more water bodies fully supports its designated aquatic life use, and thereby belongs in Category 1 or 2 of the IRG?   (Here "an affirmative determination of full support" is intended to be distinguished from simply determining that available information does not justify concluding that aquatic life use is NOT supported, which would call for putting the water body in Category 3 of the IRG, as to aquatic life use.)

If yes, would changes to the type, amount, or quality of biological assessment information improve support to your program? (Please provide specific recommendations.)

_____YES   _____NO

If no, what changes to the type, amount, or quality of biological assessment information (in the absence of evidence of failure to meet one or more chemical or physical water quality criteria) might lead to use of biological assessments as the basis for declaring a water to be fully supportive of its designated aquatic life use?

WORKSHEET FOR TOPIC 2: SUPPORT FOR CWA SECTION 303(D) AND TMDL PROGRAMS (page 4)

4.   Does the biological assessment program provide data and information used in support of stressor identification analyses for waters identified as having impaired aquatic life use based on biological assessments? If yes, were any individual (e.g., a particular pollutant or altered flow) stressors identified? (Please list them.)

_____YES   _____NO

If yes, were there any individual stressors for which biological assessment data was the sole basis of identifying the stressors? (Please list these stressors.)

If there were no individual stressors identified using only biological assessment data:

- How was biological assessment data used to supplement other kinds of data and information in the course of identifying individual stressors? (If possible, answer on a stressor-by-stressor basis)
- What, if any, categories of stressors (e.g., heavy metals, PAHs, nutrients) were identified using biological assessment data alone?

Would changes to the type, amount, or quality of biological assessment information and/or the timing of data availability provide better support for stressor identification?

_____YES   _____NO

If so, please provide specific recommendations on improvements to the biological assessment program that would improve particular aspects of your stressor identification efforts.

WORKSHEET FORTOPIC 2: SUPPORT FOR CWA SECTION 303(D) AND TMDL PROGRAMS (page 5)

5.   Does the biological assessment program provide data and information to support development of TMDLs?

_____YES   _____NO

If yes, in which of the following aspects of TMDL development have biological assessment data played a direct role?

___ Calculating of the overall water body-pollutant loading capacity:

___ Selecting a margin of safety:

___ Identifying sources of the pollutant of concern:

___ Allocating loads among existing and future sources:

___ Other aspects:

For any of these aspects, what changes to the type, amount, or quality of biological assessment information and/or the timing of data availability would enable such information to play a larger role? (If possible, answer on a TMDL function-by-function basis).

If no, what changes to the type, amount, or quality of biological assessment information and/or the timing of data availability would enable such information to play a direct role in TMDL development? (If possible, answer on a TMDL function-by-function basis.)

WORKSHEET FOR TOPIC 3: SUPPORT FOR CWA SECTION 402 NPDES PROGRAM

1.   Is biological assessment information used to support the CWA section 402 NPDES program?

_____YES_____NO

If yes, how is the NPDES program supported by biological assessment information?

Impact assessment	_____YES_____NO
Water quality-based effluent limits (WQBELs)	_____YES_____NO
Mixing zone determination	_____YES_____NO
WET limits and monitoring	_____YES_____NO
Causal diagnosis	_____YES_____NO
Other (please specify)	_____YES_____NO

Would changes to the type, amount, or quality of biological assessment information and/or the timing of data availability improve support to your program? (Please provide specific recommendations.)

If no to any of the above questions, what additional type, amount, or quality of technical information would be useful? Would changes to data collection and analysis and/or internal communication contribute to the use of biological assessments? Are there additional recommendations?

122

WORKSHEET FOR TOPIC 3: SUPPORT FOR CWA SECTION 402 NPDES PROGRAM (page 2)

2.   During NPDES permit reissuance, is information about biological condition downstream of the point source reviewed for evidence of any need to evaluate and potentially change permit limits to address observed problems? If yes, does the biological assessment program provide data and information to support the NPDES program for this purpose?

_____YES_____NO

If yes, would changes to the type, amount or quality of biological assessment information and/or the timing of data availability improve support to your program? (Please provide specific recommendations.)

If no, what additional type, amount, or quality of technical information would be useful? Would changes to data collection, data analysis, and/or internal communication (e.g., notification of permit reissuance schedule) contribute to the use of biological assessments? Are there additional recommendations?

WORKSHEET FOR TOPIC 4: SUPPORT FOR CWA SECTION 319 PROGRAM

1.   Does the biological assessment program provide data and information to support implementation of the CWA section 319 program?

_____YES   _____NO

If yes, would changes to the type, amount, or quality of biological assessment information and/or the timing of data availability improve support to the program? (Please provide specific recommendations.)

If not, what additional type, amount, or quality of technical information would be useful? Would changes to data collection and analysis and/or internal communication contribute to the use of biological assessments? Are there additional recommendations?

WORKSHEET FOR TOPIC 5: SUPPORT FOR SECTION 401 CERTIFICATION

1.   Does the biological assessment program provide data and information to support your agency's section 401 certification program?

_____YES   _____NO

If yes, would changes to the type, amount, or quality of biological assessment information and/or the timing of data availability improve support to the program? (Please provide specific recommendations.)

If not, what additional type, amount, or quality of technical information would be useful? Would changes to data collection and analysis and/or internal communication contribute to the use of biological assessments? Are there additional recommendations?

WORKSHEET FOR TOPIC 6: SUPPORT FOR [insert program]

1.   Does the biological assessment program provide data and information to support implementation of _____?

_____YES   _____NO

If yes, would changes to the type, amount, or quality of biological assessment information and/or the timing of data availability improve support to the program? (Please provide specific recommendations.)

If not, what additional type, amount, or quality of technical information would be useful? Would changes to data collection and analysis and/or internal communication contribute to the use of biological assessments? Are there additional recommendations?

SUMMARY WORKSHEET: SELF ASSESSMENT 2

Discussion Topics	YES	NO
1. Water Quality Standards		
2. CWA section 303(d) and TMDL Programs		
3. CWA section 402 NPDES Programs		
4. CWA section 319 NPS Programs		
5. CWA section 401 certification		
6.		
7.		
8.		

Summary observations and key recommendations:

APPENDIX D: TECHNICAL MEMORANDUM TEMPLATE

TECHNICAL MEMORANDUM

Technical Elements Evaluation of the [State/Tribal] Biological Assessment Program

[State/Tribal Agency]

[Location]

[Dates of Third Party Assessment]

Purpose:

To evaluate the technical program and to make recommendations for enhancements relative to design, methodology, and execution for credible data as a basis of making informed decisions regarding the ecological condition of [state/tribal agency's] surface waters.

Attendance:

Agency Participant Contact, Organization, (email)   Phone Number (XXX) (XXX-XXXX)

[List all state/tribal agency and U.S. Environmental Protection Agency (EPA) attendees]

Basis for Evaluation

Since 1990, EPA has supported the development of water quality agency biological assessment programs via the production of methods documents, case studies, regional workshops, and evaluations of individual water quality agency programs. EPA recommends that states and tribes use biological assessments to more precisely define their designated aquatic life uses and adopt numeric biological criteria necessary to protect those uses (USEPA 1990, 1991).

Overview and Summary of [State/Tribal Agency] Program and Significant Issues

The [date of evaluation] evaluation of the [state/tribal agency] biological assessment program addressed a range of topics, as summarized below. A biological program review was also completed using a standardized checklist and scoring methodology. The results are discussed as part of this memorandum.

Please provide a detailed summary of the agency's program for the flowing topics:

A. Monitoring and Assessment Program

B. Water Quality Standards (WQS): Designated Uses

_____

C. Delineation of Impaired Waters

_____

Biological assessment program evaluation

The following is a description of the current status of the program and the results of the technical elements evaluation.

Biological assessment program description

Please provide a detailed summary of the state's biological assessment program.

Critical elements evaluation

A biological program review was conducted by proceeding through the technical elements checklist (Appendix E) in accordance with the methodology described in *The Biological Program Review: Assessing Level of Technical Rigor to Support Water Quality Management* (EPA 820-R-13-001. The document includes a description of 13 technical elements of a biological assessment program, the checklist for evaluating the level of technical development for each element, and a method for characterizing the overall level of program rigor. The [water quality agency] critical elements evaluation yielded a raw score of __ out of a maximum possible score of 52. This is a Level __ program (range __ – __). The critical technical elements of biological assessment programs are described and divided into four general levels of technical development with Level 4 the highest level of rigor. A Level 4 program is able to provide the most comprehensive support for a water quality management program. As a technical program is improved, biological assessment information can be used with increasing confidence to support multiple water quality program needs for information. These needs include more precisely defined aquatic life uses and approaches for deriving biological criteria, supporting causal analysis, and developing stressor-response relationships.

Highlights of each element are indicated in Table D-1 (hypothetical example shown). The improvements that are needed to elevate the score for each element are described by element in the same order that they appear in the attached checklist as follows:

Table D-1. Example review results: The following recommendations were made to a state water quality agency as a result of their critical elements evaluation

Element	Comment
Element 1: Index Period Score assigned = 2.0	The score of 2.0 reflects a varied adherence to a seasonal index period. Logistical bottlenecks seem to be the principal reason for deviations that can extend into the following spring of each year. Elevating the score for this element will require a strict adherence to the August 15–November 15 index period.
Element 2: Spatial Sampling Design Score assigned = 2.0	The score of 2.0 conservatively reflects the synoptic design and spatial density of sampling sites that is employed. Elevating the score to the maximum of 4.0 will require a greater spatial density within watershed assessment units particularly getting beyond the "pour point" as the only sampling site on a river or stream.
Element 3: Natural Variability Score assigned = 2.0	The CE score of 2.0 should be elevated to 4.0 with the developments that are already underway including the addition of new regional reference sites and the fuller inclusion of the other bioregions.
Element 4: Reference Site Selection Score assigned = 3.0	As criteria are further refined (site-scoring process) for reference sites, the CE score of 3.0 should improve to 4.0 because it is being employed in the selection of new regional reference sites.
Element 5: Reference Conditions Score assigned = 3.0	The CE score of 3.0 should improve to 4.0 with the additional regional reference sites that are being established as part of the ongoing improvements described for elements 3 and 4.
Element 6: Taxa and Taxonomic Resolution Score assigned = 3.0	The CE score of 3.0 reflects the full development of the macroinvertebrate assemblage and the in progress development of a second and third assemblage. Reaching the CE score of 4.0 is contingent on the full development and use of a second assemblage.
Element 7: Sample Collection Score assigned = 3.0	The CE score of 3.0 reflects the full development of the macroinvertebrate assemblage (i.e., for the mountain region only) and the in-progress development of a second and third assemblage. Reaching the CE score of 4.0 is contingent on the full development and use of a second assemblage and for all applicable bioregions.

Element	Comment
Element 8: Sample Processing Score assigned = 3.0	The CE score of 3.0 reflects the full development of the macroinvertebrate assemblage for the mountain bioregion and the in progress development of the other bioregions and a second and third assemblage. Reaching the CE score of 4.0 is contingent on the full development and use of a second assemblage.
Element 9: Data Management Score assigned = 3.0	The CE score of 3.0 can be improved to 4.0 once the data management system includes all data (i.e., habitat and fish) and is readily accessible.
Element 10: Ecological Attributes Score assigned = 2.0	The CE score of 2.0 should increase with the development of the macroinvertebrate multimetric index (MMI) for all bioregions. A descriptive analysis of the biological condition gradient (BCG) for each representative bioregion and application of these concepts to the full development of the biological indicators and assemblages will improve the score to 4.0.
Element 11: Discriminatory Capacity Score assigned = 2.0	The CE score of 2.0 will be increased to at least 3.0 with the full development of the macroinvertebrate MMI and the derivation of appropriately detailed numeric biological criteria. Achieving a score of 4.0 will require that this be accomplished for a second biological assemblage.
Element 12: Stressor Association Score assigned = 2.0	The comparatively low CE score of 2.0 is a common characteristic of biological assessment programs that are in development and/or which have singularly been focused on status assessments with no or limited coordination with other environmental assessments. Improving the score for this element will occur as a result of addressing preceding elements 2, 3, 6, 10, and 11 and gaining a familiarity with how diagnostic capacity is developed. This will require some dedication to exploratory analyses in which the response of the biological assemblages is evaluated along the stressor axis of the BCG.
Element 13: Professional Review Score assigned = 2.0	The CE score of 2.0 can be elevated to 4.0 by instituting a more formal peer review process and by publishing some of the ongoing developments in peer reviewed journals.

Critical Elements Summary

Please provide a detailed summary of the agency's critical elements performance and include a discussion of ongoing program improvements that will increase the rigor of the agency's biological assessment program.

Recommendations

Summary of recommendations to the agency on how to improve the rigor of its biological assessment program and recommendations for program enhancements to support more comprehensive and efficient use of biological assessments in an agency's water quality program.

Citations

USEPA (U.S. Environmental Protection Agency). 1990. *Biological Criteria: National Program for Surface Waters.* EPA 440-5-90-004. U.S. Environmental Protection Agency, Office of Water, Washington, DC. <http://www.epa.gov/bioindicators/pdf/EPA-440-5-90-004Biologicalcriterianationalprogramguidanceforsurfacewaters.pdf>. Accessed October 2012.

USEPA (U.S. Environmental Protection Agency). 1991. *Policy on the Use of Biological Assessments and Criteria in the Water Quality Program.* U.S. Environmental Protection Agency, Washington, DC. <http://water.epa.gov/scitech/swguidance/standards/criteria/aqlife/biocriteria/upload/2002_10_24_npdes_pubs_owm0296.pdf>. Accessed February 2013.

APPENDIX E. TECHNICAL ELEMENTS CHECKLIST

The following is a checklist for evaluating the degree of development for each technical element of a biological assessment program and associated comments on the elements for the [water quality agency] biological assessment program. The point scale for each element ranges from lowest to highest resolution.

Element 1	(Lowest) 1.0	2.0	3.0	4.0 (Highest)	Comments
Index Period	Temporal variability is not taken into account.	Sampling period established based on practices of other agencies and/or literature. Sampling outside the index is not adjusted for temporal influence.	Index period established based on *a priori* assumptions regarding temporal variability of biological community. Effects of the use of index period are documented. Data collected outside the index period data might be adjusted to correct for temporal influences.	Temporal variability is fully characterized and taken into account for all data. Agency information needs and index periods are coordinated so that adherence to an index period is strict.	
					Points
					—

Element 2	(Lowest) 1.0	2.0	3.0	4.0 (Highest)	Comments
Spatial Sampling Design	Study design consisting of isolated, single, fixed-point sites.	Low density fixed station design. Multiple sites are used for assessment of a water body or watershed condition. Spatial coverage suitable for general condition assessments. Non-random designs at coarse scale used (e.g., 4–8 digit hydrologic unit code [HUC]). Inference of site data to larger unit of assessment based on "rules of thumb" and might be supplemented by upstream/downstream assessments.	Low density random or stratified random sampling design which allows for a statistically valid inference of biological condition to a spatial unit larger than a site. The primary goal is to assess aggregate condition and trends on a statewide or regional basis.	High density (e.g., intensive) monitoring at comprehensive spatial sampling design suitable for watershed assessments (e.g., 10–12 digit HUC) and in support of multiple water quality management program needs for information (e.g., condition assessments, use refinement, use attainability analyses [UAAs], permits). As needed, the spatial sampling combines monitoring designs to optimize cost and efficiency in data collection and analysis (e.g., combination of upstream-downstream, intensive, probabilistic, and/or pollution gradient designs). Typically includes a rotating sequence of watershed units organized to provide data for management program support.	**Points** —

Element 3	(Lowest) 1.0	2.0	3.0	4.0 (Highest)	Comments
Natural Variability	No or minimal partitioning of natural variability in aquatic ecosystems. Does not incorporate differences in watershed characteristics such as size, gradient, temperature, elevation, etc.	Classification scheme is based on assumed, first-order classification classes. These include strata such as fishery-based cold or warmwater classes. There is no formal consideration of regional strata such as bioregions or aggregated ecoregions. Intra-regional strata such as watershed size, gradient, elevation, temperature are not addressed. Usually applied uniformly on a statewide basis.	A fully partitioned and stratified classification scheme or modeling approach is employed. Classes and/or continuous models are defined to take critical details of spatial variability into account. Inter-regional landscape features and phenomena are appropriately sequenced with intra-regional strata. Subcategories of lotic ecotypes are defined (e.g., includes the full strata of lotic water body types). Characterization of spatial variability is confined within jurisdictional boundaries.	Scheme to fully account for natural variation is periodically refined and updated as new data and methods become available. Classes, continuous models, or both, are examined to identify the most appropriate scheme for monitoring and assessment, regulatory support, and cost-effectiveness. Developed at scales that transcend jurisdictional boundaries when necessary to strengthen inter-regional classification outcomes; recognizes the full zoogeographical aspects of biological assemblages.	**Points** —

135

Element 4	(Lowest) 1.0	2.0	3.0	4.0 (Highest)	Comments
Reference Sites Selection	Informal best professional judgment (BPJ) used in selection of control sites. No screens are used. Limited, if any, documentation and supporting rationale.	Based on "best biology" (i.e., BPJ on what the best biology is in the best water body). Minimal non-biological data used. Minimal documentation.	Selection based on narrative descriptions of non-biological characteristics. Combines BPJ with narrative description of land use and site characteristics. Might use chemical and physical data thresholds as primary filters.	Based on quantitative descriptions of non-biological characteristics with primary reliance on abiotic data on landscape conditions and land use. Chemical and physical data might be used as secondary filters or in a hybrid approach for severely altered landscapes. Independent data set used for validation.	**Points** —

Element 5	(Lowest) 1.0	2.0	3.0	4.0 (Highest)	Comments
Reference Conditions	No reference condition has been developed. Biological data are assessed using BPJ or based on the presence of targeted or iconic taxa.	Reference condition based on biology of an estimated 'best' site or water body. Single reference sites are used to assess biological data collected throughout a watershed. A site-specific control or paired watershed approach might be used.	Reference condition is based on a regional aggregate of reference site information. Data representing most of the major natural environmental gradients but limited in number and/or spatial density. Overall number and coverage of reference sites insufficient to support statistical evaluation of the biological condition at test sites.	Reference condition is based on data from many reference sites that span all major natural environmental gradients in the study area. Reference condition can be estimated for individual sites by modeling biota-environmental relationships. The number of reference sites is sufficient to support statistical evaluation of biological condition at test sites. Reference sites are resampled periodically. In highly altered regions or water body types, alternative methods are used to develop reference condition.	**Points** —

Element 6	(Lowest) 1.0	2.0	3.0	4.0 (Highest)	Comments
Taxa and Taxonomic Resolution	One taxonomic assemblage (e.g., benthic macroinvertebrates, fish, algae, aquatic macrophytes). Very coarse taxonomic resolution (e.g., order/family). Expertise: amateur naturalist or stream watcher. Validation: none. QA/QC: none.	One taxonomic assemblage. Low taxonomic resolution (e.g., family). Expertise: novice or apprentice biologist. Validation: family level certification for macroinvertebrates. No certification available for fish or algae. QA/QC: mostly for taxonomic confirmation of voucher collections. Some sorting QA/QC implemented.	One taxonomic assemblage. Fine taxonomic resolution: genus/species for benthic macroinvertebrates and algae, species for fish. Expertise: trained taxonomist. Validation: genus-level certification or equivalent for benthic macroinvertebrates. Expert fish taxonomist or equivalent. Formal courses or training in algal taxonomy. QA/QC: addresses measuring bias, precision, and accuracy in all phases of sample processing through identification (e.g., outside validation of identification); voucher collection maintained.	Same as Level 3 except that two or more taxonomic assemblages are assessed. Rationale for selection of taxonomic groups should be well documented.	**Points** —

Element 7	(Lowest) 1.0	2.0	3.0	4.0 (Highest)	Comments
Sample Collection	Approach is cursory and relies on operator skill and BPJ. Training limited to that which is conducted annually for non-biologists who compose the majority of the sampling crew. Methods are not systematically documented as standard operating procedures (SOPs).	Textbook methods are used without considering the applicability of the methods to the study area. SOPs to specify methods but methods are neither well documented nor evaluated for producing comparable data across agencies. A cursory QA/QC document might be in place. Training consists of short courses (1–2 days) and is provided for new staff and periodically for all staff.	Methods are evaluated for applicability to study area and refined (if needed). Detailed and well documented SOPs are updated periodically and supported by in-house testing and development. A formal QA/QC program is in place with field replication requirements. Rigorous training required for all professional staff.	Same as Level 3, but methods cover multiple assemblages. A field audit of sampling crews is performed annually to ensure that protocols and proper sample handling/documentation are followed.	
					Points
					—

Element 8	(Lowest) 1.0	2.0	3.0	4.0 (Highest)	Comments
Sample Processing	Organisms are sorted, identified, and counted in the field using dichotomous keys.	Organisms are sorted, identified, and counted primarily in the field by trained staff. Adequate QA/QC is not possible. For fish, cursory examination of presence and absence only. Agency SOPs not developed or published.	All samples (except for fish) are processed in the laboratory. A formal QA/QC program is in place. Rigorous training is provided. Voucher organisms are retained for ID verification. SOPs are published and available to others.	Same as Level 3, but applied to multiple assemblages. Subsampling level is tested. Presence of fish deformities, erosions, lesions, tumors (DELT) and other anomalies are quantified and documented.	
					Points
					—

Element 9	(Lowest) 1.0	2.0	3.0	4.0 (Highest)	Comments
Data Management	Sampling event data organized in a series of spreadsheets (e.g., by year, by data-type). QA/QC is cursory and mostly for transcription errors. Might be paper files only.	Databases for physical-chemical, and biological data, and geographic information exist (Access, dBase, Geographic Information System [GIS], etc.) but are not linked or integrated. Data-handling methods manuals are available. QA/QC for data entry, value ranges, and site locations. A documented data dictionary defines data fields in terms of field methods and data collection.	Relational databases that integrate all biological, physical, and chemical data (Oracle, SQL Server, Access, etc.). Validation checks that guard against inadvertently storing incorrect or incomplete sampling data. Fully documented and implemented QA/QC process. Structure provides for data export and analysis via query includes dedicated database management. Fully documented data dictionary. Access to all databases is available for routine analysis in support of condition assessment.	Same as Level 3 adding automated data review and validation tools. Numerous built-in data management and analysis tools to support routine and exploratory analyses. Ability to track history of changes made to the data. Ability to control who has privilege to change, update, or delete data. Data import and export tools. Integrated connection to GIS showing monitored sites in relation to other relevant spatial data layers. Fully documented metadata according to accepted database standards. Reports on commonly used endpoints are easily retrieved (e.g., menu driven).	**Points** —

Element 10	(Lowest) 1.0	2.0	3.0	4.0 (Highest)	Comments
Ecological Attributes	Biological program relies solely on the evaluation of the presence or absence of targeted or key species. No rationale is provided for selection of indicators. Assessment endpoints and ecological attributes are not defined.	Biological program based on "off the shelf" indicators for one biological assemblage. Rationale for selection of indicators is partially documented. Generic assessment endpoints and ecological attributes are defined but not specifically evaluated for state or regional conditions.	Biological program based on well-developed ecological attributes for one biological assemblage. Rationale for attribute selection is thorough and well-documented. Explicit linkage is provided between management goal, assessment endpoints, and ecological attributes.	Same as Level 3, but biological program based on well-developed ecological attributes for two or more biological assemblages (e.g., faunal, flora) for more complete assessment of the members of an aquatic community.	**Points** —

Element 11	(Lowest) 1.0	2.0	3.0	4.0 (Highest)	Comments
Discriminatory Capacity	Coarse method (low signal) and detects only high and low values. Supports distinguishing only extreme change in biological condition at the upper and lower ends of a generalized stress gradient.	A biological index for one assemblage is established but is not calibrated for water body classes, regional or statewide applications. BPJ based on single dimension attributes. The index can distinguish two general levels of change in biological condition along a generalized stress gradient.	A biological index for one assemblage has been developed and calibrated for statewide or regional application and for all classes and strata of a given water body type. The index can distinguish 3 to 4 increments of biological change along a continuous stress gradient. Supports narrative evaluations (e.g., good, fair, poor) based on multimetric or multivariate analyses that are relevant to the selected ecological attributes (Technical Element 10).	Same as Level 3 but biological indices for two or more assemblages have been developed and calibrated. Additionally, the indices can distinguish finer increments of biological change along a continuous stress gradient. The number of increments that potentially can be distinguished is dependent on water body type and natural climatic and geographic factors.	**Points** —

Element 12	(Lowest) 1.0	2.0	3.0	4.0 (Highest)	Comments
Stressor Association	No ability to develop relationships between biological responses and anthropogenic stress.	Site-specific paired biological and stressor samples for studies of an individual water body or a segment of a water body (e.g., a stream reach). Stress-response relationships are developed based on assemblage attributes at coarse level taxonomy (e.g., family for benthic macroinvertebrates). Information might be used on a case-by-case basis to inform a first order causal analysis.	Low spatial resolution for paired biological and stressor samples in time and space across the state at basin or sub-basin scale (e.g., HUC 4–8). Stress-response relationships developed for one assemblage using regression analysis. Taxonomy at level sufficient to detect patterns of response to stress (e.g., species or genus for benthic macroinvertebrates or periphyton, species for fish). Relational database supports basic queries. Information is frequently used to inform causal analysis. Reevaluation of stress-response relationships on an as-needed basis.	High spatial resolution for paired biological (including DELT anomalies and other indicators of organism health) and stressor samples in time and space across the state at watershed or subwatershed scales (e.g., HUC 10–12). Other data (e.g., watershed characteristics, land use data and information, flow regime, habitat, climatic data) are linked to field data for source identification. Stress-response relationships are fully developed for two or more assemblages, stressors, and their sources using a suite of analytical approaches (e.g., multiple regression, multivariate techniques). Relational database supports complex queries. Information is routinely used to inform causal analysis and criteria development. Ongoing evaluation of stress-response relationships and monitoring for new stressors is supported.	Points —

143

Element 13	(Lowest) 1.0	2.0	3.0	4.0 (Highest)	Comments
Professional Review	Review is limited to editorial aspects. No technical review.	Internal technical review only.	Outside review of documentation and reports are conducted on an ad hoc basis.	Formal process for technical review to include multiple reference and documented system for reconciliation of comments and issues. Process results in methods and reporting improvements. Can include production of peer-reviewed journal publications by the agency.	**Points** —

EPA 820-R-13-001

www.ingramcontent.com/pod-product-compliance
Lightning Source LLC
Chambersburg PA
CBHW051512170526
45166CB00001B/490